Edexcel
GCSE MATHEMATICS

PRACTICE BOOK: Foundation

Gareth Cole   Peter Jolly   David Kent   Keith Pledger

Note: There are no practice exercises for units 12 and 24.

Edexcel
Success through qualifications

Heinemann

WITHDRAWN

D0314069

About this book

This book provides a substantial bank of additional exercises to complement those in the Edexcel GCSE Mathematics course textbook and offers a firm foundation for a programme of consolidation and homework.

Extra exercises are included for every topic covered in the course textbook, with the exception of the Using and applying mathematics and the Calculators and computers units.

Clear links to the course textbook exercises help you plan your use of the book:

Exercise 1.1	Links: (*1A, B*) 1A, B

This exercise is linked to exercises 1A and 1B in the old edition of the course textbook.

This exercise is linked to exercises 1A and 1B in the new edition of the course textbook.

Please note that the answers to the questions are provided in a separate booklet, available free when you order a pack of 10 practice books. You can buy further copies direct from Heinemann Customer Services.

Also available from Heinemann:

Edexcel GCSE Mathematics: Foundation

The Foundation textbook provides a complete course for the Intermediate tier examination.

Revise for Edexcel GCSE: Foundation

The Foundation revision book provides a structured approach to pre-exam revision and helps target areas for review.

1 Number

1 Draw a place value diagram and write in:
   (a) the number 5348
   (b) a four figure number with a 7 in the Thousands column
   (c) a three figure number with a 4 in the Hundreds column
   (d) a two figure number with a 5 in the Tens column
   (e) a six figure number with a 4 in each column
   (f) a five figure number with a 5 and a 6 in alternate columns
   (g) a three figure number with a 4 in the first and last columns
   (h) a four figure number with a 9 in the first and last columns
   (i) a five figure number with a 6 in the left hand column and a
       zero in the Units column
   (j) a four figure number with a 7 in the Units column and digits
       less than 7 in all other columns

2 Write these numbers in words:
   (a) 47   (b) 73   (c) 128   (d) 207   (e) 1300   (f) 4236   (g) 32 406

3 Use the place value diagram to help you write these numbers in
   words:
   (a) 4563   (b) 64 500   (c) 302 709   (d) 6 538 741
   (e) 13 773 406   (f) 45 000 000   (g) 23 500 004

4 (a) The distance from the Earth to the Moon is 400 000
       kilometres. Write this number in words.
   (b) The distance from the Earth to the Sun is ninety three
       million miles. Write this number in figures.

5 The attendance figures at six major sporting events are listed
   below.

   Boxing Match : Five Thousand   Horse Race : 120 000
   Soccer Match : 42 000   Snooker : 850
   Hockey International : Seven Thousand   Rugby Cup Final : 80 000

   List these attendance figures in order. Start with the smallest.

6 Put the numbers in the cloud in order.
   Start with the smallest.

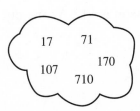

17   71   107   170   710

Exercise 1.2   Links: (*1C–G*) 1C–G

1 Draw number lines going from 0 to 20 to help you with each of these:
   (a) increase 7 by 5   (b) decrease 13 by 4
   (c) decrease 5 by 5   (d) increase 2 by 17
   (e) decrease 20 by 3   (f) increase 14 by 6
   (g) Work out what increase moves 7 to 15.
   (h) Work out what decrease moves 16 to 5.
   (i) What change moves 8 to 14?
   (j) What change moves 13 to 2?
   (k) What change moves 19 to 1?

2 Work out:
   (a) the total of 17 and 46   (b) 68 plus 32   (c) $427 + 86$

3 A coach has 54 seats downstairs and 38 seats upstairs. How many seats does the coach have altogether?

4 Work out the sum of all the whole numbers from 10 to 20.

5 James has 48 CDs. Angela has 25 CDs. How many CDs more than Angela does James have?

6 Work out:
   (a) the difference between 200 and 86   (b) $3000 - 238$
   (c) the number which is 17 less than 1001   (d) 320 minus 18

7 When it was new the value of a car was £16 580. When it was five years old the value of this car was £4600. Work out the difference between the value of the car when it was new and the value of the car when it was five years old.

8 Paddy will have his 45th birthday in the year 2027. In which year was Paddy born?

9 Work out:
   (a) the product of 13 and 7   (b) 48 multiplied by 17
   (c) 23 times 25   (d) 54×26
   (e) 15 multiplied by itself   (f) $2 \times 8 \times 20$

10 The cost of a piano lesson is £12. Work out the total cost of:
   (a) 10 piano lessons
   (b) 35 piano lessons
   (c) 200 piano lessons

11 A group of 8 friends go to the cinema. Each cinema ticket costs £6.95. Work out the total cost of the 8 cinema tickets.

12 Jennifer has written a geography project. The project is 32 pages long. She needs to make a photocopy. The cost of making a photocopy is 4 p for each page. Work out the total cost of making the photocopy of all 32 pages.

13 Work out:
   (a) $325 \div 25$ **(b)** 800 divided by 40
   (c) $\frac{720}{18}$ **(d)** the number of 20 p coins in £6

14 Mr Khan won £720 in the Lottery. He shared this money equally between himself, his wife and his two children. Work out how much each person received.

15 Martin has 360 bottles of wine. He is going to put them in crates. Each crate holds 12 bottles. How many crates will Martin need to use?

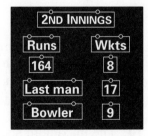

Exercise 1.3   Links: (*1G, H, I*) 1H, I, J

1 **(a)** Lucea Cricket Club scored 205 runs in their first innings and 175 runs in their second innings. How many runs did they score altogether?
   (b) In the same match, Nayland Cricket Club scored a total of 345 runs. The team which wins the match is the team which scores the most runs.
   (i) Which team won the match   **(ii)** and by how many runs?

2 Joan earns £4 for each hour she works. Last week she earned £148. How many hours did Joan work last week?

3 Steven was born on 31st October 1971. Work out the date of his 30th birthday.

4 A single entered the charts at number 16. The following week it rose to number 9.
   (a) By how many places did the single rise in this week?

   In its third week in the charts the single fell 4 places.
   (b) At what position in the charts was the single then?

5 Fatima bought 5 books and 3 pens. She paid £6.50 for each book and £1.05 for each pen. Work out the total amount Fatima paid for the 5 books and 3 pens.

6 Maureen bought a box of chocolates. The box contained 24 chocolates. Maureen ate six of the chocolates herself. She then shared the rest equally between her 3 cousins. How many chocolates did each cousin receive?

7 Round each of these numbers:
  (a) 364 to the nearest ten
  (b) 4872 to the nearest thousand
  (c) 25 638 to the nearest hundred
  (d) 324 561 to the nearest 10

8 Last Monday 43 617 commuters arrived at Liverpool Street Station. Round this number of commuters to:
  (a) the nearest 10
  (b) the nearest 100
  (c) the nearest 1000

9 Write each of these numbers to 1 significant figure:
  (a) 14   **(b)** 25   **(c)** 37   **(d)** 146
  (e) 231   **(f)** 407   **(g)** 4870   **(h)** 3714
  (i) 1758   **(j)** 6531   **(k)** 10 6317 **(l)** 22 493

10 The maximum speed of the train is 142 mph. Write this number to 1 significant figure.

11 The distance Jane travelled was 1865 miles. Write this number to 1 significant figure.

12 Showing all your rounding, make an estimate of the answer to:
  (a) $\dfrac{62 \times 49}{32}$ **(b)** $\dfrac{317 \times 402}{198}$ **(c)** $\dfrac{3008 \times 21}{17 \times 43}$

13 A theatre has 1205 seats when it is full. A ticket for a seat at the theatre costs £6.95. Last Saturday all of the tickets for the seats were sold. Estimate the total cost of all these tickets.

Exercise 1.4   Links (*1J, K, L*) 1K, L, M, N

1 (a) Write down all the even numbers from the list
       4, 9, 20, 56, 73, 115, 906, 1002, 43 278
   (b) Write down all the odd numbers from the list
       4, 9, 23, 56, 74, 218, 827, 2001, 56 983

2 (a) Write down all the prime numbers between 20 and 30.
   (b) Write down the smallest even number greater than 50.
   (c) Write down the largest odd number less than 50.
   (d) Write down the smallest prime number greater than 50.

3 Write down all the factors of:
   (a) 8   (b) 24   (c) 100   (d) 280

4 Find all four factors of the number 111.

5 Write down the first 5 multiples of:
   (a) 3   (b) 10   (c) 12   (d) 200

6 Find a number which is a multiple of 5 and also a multiple of 8.

7 Find the first multiple of 9 which is greater than 50.

8 Find the number which is a factor of:
   (a) 12 and 15   (b) 14 and 63
   (c) 15 and 50   (d) 33 and 77

9 Find both of the numbers which are factors of:
   (a) 20 and 50   (b) 42 and 105

10 Copy these three rows of numbers.

       1 2 3 4 5 6 7 8 9 10
      11 12 13 14 15 16 17 18 19 20
      21 22 23 24 25 26 27 28 29 30

   (a) Draw a ○ around each multiple of 3.
   (b) Draw a X through each multiple of 4.
   (c) Write down all the numbers less than 30 which are both a
       multiple of 3 and a multiple of 4.
   (d) Find the first number greater than 30 which is a multiple of
       both 3 and 4.

11 The first three square numbers are shown as dot patterns.
   (a) Draw the dot pattern for the 4th square number.
   (b) Work out the 4th square number.
   (c) Work out the 5th square number.

   1st = 1   2nd = 4   3rd = 9

12 Make a list of all the square numbers from the first to the twelfth.

13 Work out:
    (a) 6^2 **(b)** 10^2 **(c)** 20^2 **(d)** 25^2

14 Work out:
    (a) $\sqrt{25}$ **(b)** $\sqrt{64}$ **(c)** $\sqrt{100}$ **(d)** $\sqrt{400}$

15 Make a list of all the cube numbers from the first to the twelfth.

16 Work out:
    (a) 5^3 **(b)** 10^3 **(c)** 20^3 **(d)** 25^3

17 Find a number between 50 and 70 which is both a square number and a cube number.

18 Find the value of:
    (a) 7 squared   **(b)** 2 cubed
    (c) 10 to the power 5   **(d)** 10 to the power 8
    (e) 4^3

19 Find the value of:
    (a) 5^2 **(b)** 4^3 **(c)** 7^2 **(d)** 7^3
    (e) 1^4 **(f)** 3^3 **(g)** 12^2 **(h)** 10^5
    (i) 1^2 **(j)** 6^3 **(k)** 4^2 **(l)** 10^9

20 Which is the biggest and by how much:
2^3 or 3^2?

Exercise 1.5   Links (*1M*) 1O, P

1 Copy the number line and fill in the missing numbers.

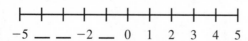

$$-5 \ \_\_ \ \_\_ \ -2 \ \_\_ \ 0 \ 1 \ 2 \ 3 \ 4 \ 5$$

2 On New Year's Day, the temperature in °C, in five cities were

    London   Glasgow   Moscow   Athens   Madrid
      0°   −2°   −14°   12°   8°

Write these temperatures in order, starting with the lowest.

3 Find the missing two numbers in each of the sequences:
    (a) 12, 7, 2, —, —, −13
    (b) −16, −12, −8, —, —, 4, 8

4 Find the number which is:
 (a) 7 less than 4   **(b)** 4 less than 0
 (c) 8 more than -3 **(d)** 10 more than -10
 (e) 1 less than -5 **(f)** 12 more than -5
 (g) 6 less than -1 **(h)** 20 more than -5
 (i) 20 less than -30 **(j)** 100 more than -50
 (k) 5 less than -100 **(l)** 200 less than 0
 (m) 14 more than -12 **(n)** 5 bigger than -4
 (o) 6 bigger than -7 **(p)** 10 smaller than -3
 (q) 150 smaller than -50 **(r)** 200 bigger than -300

5 The temperature in Belfast at 6 am on Christmas Day was $-3°C$. By noon on Christmas Day the temperature in Belfast had risen by $9°C$. Work out the temperature in Belfast at noon on Christmas Day.

6 The temperature at the top of a mountain is $-8°C$. The temperature at the bottom of this mountain is $20°C$ higher than the temperature at the top. Work out the temperature at the bottom of the mountain.

7 When an aeroplane takes off, the temperature at ground level is $15°C$. The aeroplane flies to a height of 9000 metres. At this height the temperature is $35°C$ lower than the temperature at ground level. Work out the temperature at 9000 metres.

8 The highest point of a tree is 6 metres above ground level. The lowest point of the roots of the tree is 8 metres below ground level. Work out the distance from the lowest point of the roots of the tree to the highest point of the tree.

9 The temperature at the top of a mountain is $-14°C$. The temperature at the bottom of the mountain is $12°C$. Work out the difference between the temperature at the bottom of the mountain and the temperature at the top of the mountain.

10 On the 19th December the difference between the temperatures in Moscow and Luxor was $31°C$. Luxor was warmer than Moscow.
The temperature in Moscow on the 19th December was $-10°C$.
Work out the temperature in Luxor on the 19th December.

11 Work out the values of:
 (a) $5 + -3$ **(b)** $4 + -7$ **(c)** $-2 - -6$
 (d) $-3 - 3$ **(e)** $-3 - -7$ **(f)** $-5 + -2$
 (g) -3×-2 **(h)** -5×-3 **(i)** -2×7
 (j) 3×-5 **(k)** -5×-2 **(l)** -5×-5
 (m) 5×-2 **(n)** -3×-8 **(o)** -6×4

2 Algebra 1

Exercise 2.1   Links: (*2A, B, C, D*) 2A, B, C, D

1 Use algebra to write:
   (a) 4 more than a
   (b) 3 less than b
   (c) c with 5 added
   (d) d more than c
   (e) $4e$ with 7 subtracted

2 Write these in short form:
   (a) $x + x + x$
   (b) $y + y$
   (c) $b + b + b + b + b$
   (d) $w + w + w + w$
   (e) $p + p + p + p + p + p + p + p$
   (f) $q + q + q + q + q + q$

3 Write these out in a longer form:
   (a) $3y$ **(b)** $4x$ **(c)** $5w$ **(d)** $2z$
   (e) $5a$ **(f)** $7c$ **(g)** $9d$ **(h)** $4y$

4 Make these expressions simpler by adding or subtracting like terms:
   (a) $2x + 3x$ **(b)** $3y + y$ **(c)** $5a - 2a$
   (d) $4w + 3w$ **(e)** $5a + 4a + 3a$ **(f)** $8b - 3b$
   (g) $5x + 6x - 3x$ **(h)** $2d + 4d$ **(i)** $5y - 3y + 2y$
   (j) $12x - x$ **(k)** $10a - 10a$ **(l)** $9s - 9s + s$

5 Simplify these expressions by collecting like terms:
   (a) $3y + 2x + 4y + 5x$
   (b) $6a + 2b + 5a + 5b$
   (c) $3d + 2f + 3d$
   (d) $3s + 2t - s + 2t$
   (e) $4d + d + 3d - 5d$
   (f) $5x - 2y + 3x + 2x$
   (g) $4p + 5p + 6p - 9p - 5p$
   (h) $6a + 2a - 3a + 4a - 4a$
   (i) $3x + 1 + 3x + 1 + 2$
   (j) $5 + 2a + 3a - 2a + 4$
   (k) $4a + 7b - 2a - 6b - b$
   (l) $4y - 3x + 2 + 3x - 4y + 2$
   (m) $5b - 3b + 1 - 7b - 1 - 2b$
   (n) $3p + 7p - 10p + 7 + p$
   (o) $4x + 3y + 2y + 5x - 3x + 2y$
   (p) $8a + 4a - 5a + 2a - 7a + 3a$

Exercise 2.2   Links: (*2E, F, G*) 2E, F, G

Use multiplication signs to write the expressions out in a longer form:

1 ab 2 cde 3 $2a$ 4 $3ab$
5 $5xyz$ 6 $10abc$ 7 $7mnp$ 8 $4abcd$
9 $15xyz$ 10 $9abcd$ 11 $3xyz$ 12 $21abc$

Write these expressions in a simpler form:

13 $a \times b$ 　　　　　**14** $x \times y \times z$ 　　　　　**15** $3 \times w \times s \times t$
16 $5 \times m \times n$ 　　　**17** $4 \times b \times c \times d$ 　　**18** $h \times k \times l$
19 $2 \times s \times t$ 　　　**20** $7 \times a \times b \times d$ 　　**21** $s \times 5 \times t \times w$
22 $2x \times 3y$ 　　　　　**23** $4a \times 5b$ 　　　　　**24** $5a \times 2b$
25 $a \times b \times c$ 　　　**26** $6s \times 2t$ 　　　　　**27** $8y \times 3z$
28 $2a \times 2b \times 2c$ 　　**29** $5g \times 2f$ 　　　　　**30** $9a \times 3b \times c$
31 $10x \times 9y \times 3z$ 　　**32** $5d \times 3e$ 　　　　　**33** $e \times 2f \times g$
34 $6a \times 4b \times 2c$ 　　**35** $4a \times 4b \times 4c$ 　　**36** $11a \times 10b \times 3c$

Write these expressions using powers:

37 $x \times x \times x \times x$ 　　**38** $y \times y$ 　　　　**39** $a \times a \times a \times a \times a$
40 $s \times s \times s$ 　　　　**41** $b \times b \times b \times b \times b$ 　**42** $p \times p \times p$

Write these expressions in full:

43 a^2 　　　**44** x^4 　　　**45** y^3 　　　**46** x^{10}
47 t^3 　　　**48** d^6 　　　**49** k^2 　　　**50** y^5

51 Find the value of these powers:
　(a) 2^2 　(b) 3^3 　(c) 4^4 　(d) 3^4 　(e) 7^2
　(f) 6^3 　(g) 7^3 　(h) 5^1 　(i) 3^5 　(j) 6^4

Exercise 2.3　　　　　　　　Links: (*2H–N*) 2H–N

1 Use BiDMAS to help find the value of these expressions:
　(a) $4 + (5 + 2)$ 　(b) $6 - (4 + 1)$ 　(c) $6 \times 3 + 2$
　(d) $25 \div 5 + 5$ 　(e) $11 - (4 - 3)$ 　(f) $30 \div 5 - 3$
　(g) $(4 + 5)^2$ 　(h) $2 \times (3 + 4)^2$ 　(i) $3^2 + 4^2$
　(j) $\dfrac{6^2 - 4^2}{5}$ 　(k) $\dfrac{(1 + 3)^2}{2^2 - 2}$ 　(l) $2 \times 3^2 + 3 \times 4^2$

2 Expand the brackets in these expressions:
　(a) $4(a + b)$ 　　　　　(b) $5(x - y)$
　(c) $2(3x + 2y)$ 　　　　(d) $3(2x - 3y)$
　(e) $7(2x - y + 3z)$ 　　(f) $11(3a + 9b)$
　(g) $4(x + y) + 3(x + y)$ 　(h) $5(a - b) + 6(2a + 3b)$
　(i) $2(p + 3q) + 5(2p - q)$ 　(j) $7(s + 2t) + 3(2s - 3t)$

3 Work out:
　(a) $15 - (4 + 2)$ 　(b) $18 - (5 + 3)$ 　(c) $8 - (6 - 5)$
　(d) $13 - (7 + 6)$ 　(e) $9 - (8 - 3)$ 　(f) $21 - (8 - 5)$

4 Write these expressions as simply as possible:
　(a) $3x + 4y - (x + y)$ 　　(b) $2a + 5b - (a + b)$
　(c) $5s + 4t - (3s - 2t)$ 　　(d) $3(4w + 5x) - 2(3w + 2x)$
　(e) $4(x + 2y) - 3(x + 4y)$ 　(f) $5(2a + 3b - 4c) - 3(3a + 2b - 2c)$
　(g) $3(2p - 3q) + 2(3p + 2q) - 4(2p - q)$

(h) $5(3a - 2b) + 3(2b - 3a) - 4(a + 5b)$
(i) $4(3x + y) - 3(5x - 3y) + 2x - 5y$
(j) $7(2p + 5q) - 5(3q - 4p) - 3(6p + 2q)$

Exercise 2.4   Links: (2O, P, Q) 2O, P, Q

1 Factorize:
   (a) 18 (b) 12 (c) xy
   (d) $4x$ (e) a^4 (f) $10ab$

2 Factorize:
   (a) $a^2 + 2a$ (b) $x^2 - 2xy$ (c) $x^2 - xy$
   (d) $6x + 3$ (e) $8a - 8$ (f) $5b^2 + b$
   (g) $5d + 15$ (h) $9e - 9$ (i) $2y^3 + 2y$
   (j) $8a + 24b$ (k) $3x^2 - 2x$ (l) $5ab - 3ac$

3 Simplify the following expressions:
   (a) $2(a + 5) + 3(a + 2)$ (b) $3(x + 3) + 2(x - 3)$
   (c) $4(5 - y) + 2(3y + 2)$ (d) $a(a + 2) + a(a + 3)$
   (e) $2(y + 3) - 3(2y + 1)$ (f) $x(3x - 4) - x(2x + 5)$

Exercise 2.5   Links: (2O–T) 2R–W

1 Find the two missing numbers in these number patterns. Write
   down the rule for each pattern too.
   (a) 4, 6, 8, —, —, 14, 16
   (b) 5, 8, 11, 14, —, —, 23, 26
   (c) 9, 19, —, —, 49, 59, 69
   (d) 1, 7, 13, —, —, 31, 37
   (e) 17, 21, —, —, 33, 37, 41
   (f) 29, 25, 21, —, —, 9, 5
   (g) 98, 88, 78, —, —, 48, 38
   (h) 47, 41, —, —, 23, 17, 11
   (i) 17, 15, —, —, 9, 7, 5
   (j) 80, 71, 62, —, —, 35, 26
   (k) 3, 6, 12, —, —, 96, 192
   (l) 2, 6, 18, —, —, 486, 1458
   (m) 1, 6, —, —, 1296, 7776
   (n) 6, 30, 150, —, —, 18 750
   (o) 11, 110, —, —, 110 000, 1 100 000
   (p) 64, 32, —, —, 4, 2
   (q) 729, 243, —, —, 9, 3, 1
   (r) 3072, 768, —, —, 12, 3
   (s) 300 000, 30 000, —, —, 30, 3
   (t) 12 500, 2500, —, —, 20, 4

2 Copy the pattern, find the difference and the rule for each one, and the next number.
   (a) 5, 7, 9, 11, 13, ...
   (b) 3, 8, 13, 18, 23, ...
   (c) 2, 4, 6, 10, 16, ...
   (d) 1, 3, 4, 7, 11, ...
   (e) 1, 4, 9, 16, 25, ...
   (f) 3, 3, 6, 9, 15, ...
   (g) 1, 6, 11, 16, 21, ...

3 Find the general rule for the nth term for each pattern. Then use your rule to find the 20th term.
   (a) 2, 4, 6, 8, 10, 12, ...
   (b) 4, 8, 12, 16, 20, 24, ...
   (c) 9, 18, 27, 36, 45, 54, ...
   (d) 3, 6, 9, 12, 15, 18, ...
   (e) 4, 6, 8, 10, 12, 14, ...

4

Find the general rule for the number of matches needed to make the nth pattern in this sequence.

3 Angles and turning

1 Write down which of these are turning movements:
   (a) using a door handle to open a door
   (b) pulling a bucket out of a well
   (c) winding up a clockwork toy
   (d) a stone dropped from a cliff
   (e) the minute hand of a clock as time passes
   (f) using a remote control to turn the TV on

2 Here is a map of a village.

School   Post Office   Church
  •   •   •

                                                      N
                                                      ↑
Shop   Village Hall   Bridge   W ──┼── E
  •   •   •   │
                                                      S

Village
Green   Pond   Windmill
  •   •   •

Peter is standing at the village hall. He is facing North.
   (a) After a quarter turn clockwise what is he facing?
       Which compass direction is he facing?

Again Peter is standing at the village hall. He is facing North.
   (b) After a half turn anticlockwise what is he facing?
       Which compass direction is he facing?

Mary is at the pond facing towards the shop.
   (c) What direction is she facing?
   (d) After a three-quarters turn anticlockwise what is she facing?
   (e) What direction is she facing now?

3 What direction is:
   (a) the church from the school
   (b) the village green from the school
   (c) the shop from the post office
   (d) the post office from the bridge
   (e) the windmill from the school?

4 How much does the hour hand of a clock turn between:
(a) 2 pm and 8 pm
(b) 4 pm and 10 pm
(c) 10 am and 1 pm
(d) 1 am and 10 am
(e) 3 am and 3 pm
(f) 9:30 am and 9:30 pm?

Exercise 3.2   Links: (*3B, C*) 3B, C

In each case, name the marked angle and say whether it is acute, obtuse or right-angled.

1

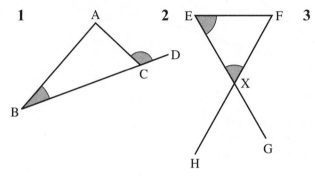

2

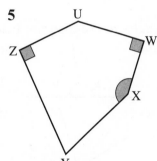

3

4

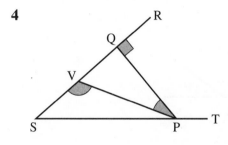

5

6 From the diagrams above, estimate the size of these angles:
(a) BAC
(b) HXG
(c) NKL
(d) NLM
(e) VSP
(f) QPT

7 Draw a quadrilateral with two obtuse angles.

Exercise 3.3   Links: (*3D–H*) 3D–H

1 Measure the angles as accurately as you can:

(a)   (b)   (c)

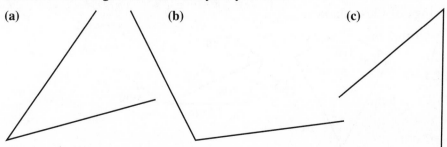

2 Measure all three angles in the following triangles:

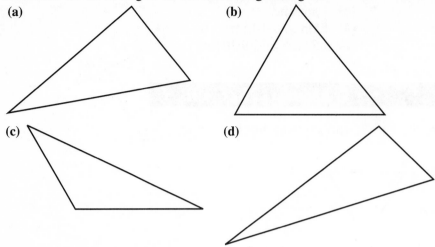

(a)   **(b)**

(c)   **(d)**

3 Draw and label these angles:
  (a) ABC = 20°   **(b)** DEF = 40°   **(c)** GHI = 60°
  (d) JKL = 35°   **(e)** MNO = 75°   **(f)** PQR = 105°
  (g) STU = 77°   **(h)** VWX = 13°   **(i)** YZA = 129°
  (j) BCD = 146°

4 Calculate the marked angles in the diagram below.

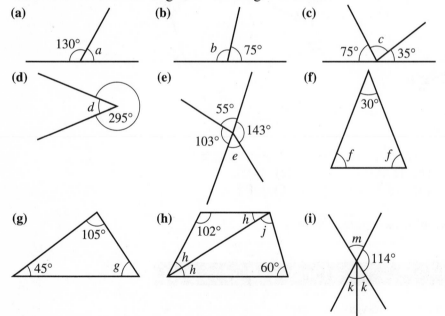

(a)   130°   a

(b)   b   75°

(c)   75°   c   35°

(d)   d   295°

(e)   55°   143°   103°   e

(f)   30°   f   f

(g)   105°   45°   g

(h)   102°   h   j   h   h   60°

(i)   m   114°   k   k

5 Make accurate drawings of these shapes.

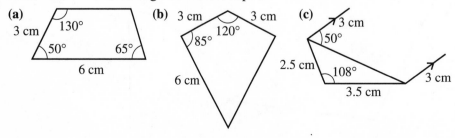

(a)   3 cm   130°   50°   65°   6 cm

(b)   3 cm   3 cm   120°   85°   6 cm

(c)   3 cm   50°   2.5 cm   108°   3.5 cm   3 cm

Exercise 3.4   Links: (3*1*) 3I

1 State the lines which are parallel to each other.

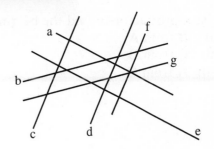

2 Copy and mark the parallel lines.

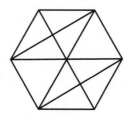

In questions **3–10** find the size of the marked angle. Give reasons for your answer.

3

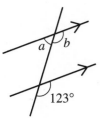

4

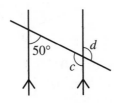

5

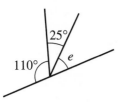

6

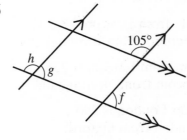

7

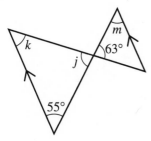

8

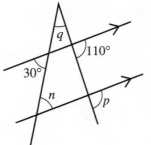

9

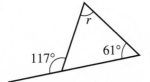

10

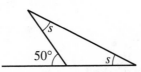

Exercise 3.5   Links: (*3J*) 3J

1 Use a protractor to find the bearings of:
   (i)   B from A
   (ii) C from B
   (iii) A from C

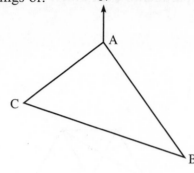

2 Find the bearings of:
   (i)   Q from R
   (ii) P from R
   (iii) R from P

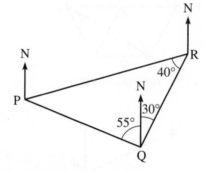

3 The bearing of Sheffield from Gloucester is 015°.
   What is the bearing of Gloucester from Sheffield?

4 The bearing of York from Leeds is 060°.
   What is the bearing of Leeds from York?

5 The bearing of Birmingham from Liverpool is 147°.
   What is the bearing of Liverpool from Birmingham?

6 The bearing of Nottingham from Lincoln is 235°.
   What is the bearing of Lincoln from Nottingham?

7 The bearing of Oxford from London is 290°.
   What is the bearing of London from Oxford?

8 Use the diagram to find the bearing of each of the towns from
   Luton.

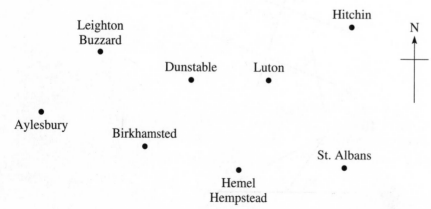

4 Fractions

1 Copy each of these shapes. In each case write down the fraction of the shape that has been shaded.

(a) (b) 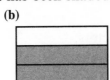 (c) (d)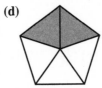

2 Make 4 copies of this shape. Shade them to show each of the fractions:

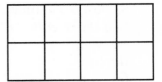

(a) $\frac{1}{4}$ (b) $\frac{3}{4}$ (c) $\frac{5}{8}$ (d) $\frac{6}{8}$

3 David travels 60 miles to work. He travels the first 20 miles by car. He travels the remainder of the distance by rail.
(a) What fraction of the journey does David travel by car?
(b) What fraction of the journey does David travel by rail?

4 Yvonne keeps a diary. According to her diary, during the 24-hour period last Tuesday, she spent 8 hours at school, 1 hour travelling, 3 hours at work, 6 hours asleep, 2 hours doing homework and 4 hours playing games at home. Write down the fraction of this 24-hour period Yvonne spent:
(a) at school (b) travelling   (c) at work
(d) asleep   (e) doing homework (f) playing games at home.

5 Change each of these improper fractions to mixed numbers:
(a) $\frac{7}{6}$ (b) $\frac{17}{6}$ (c) $\frac{22}{7}$ (d) $\frac{43}{10}$ (e) $\frac{28}{24}$ (f) $\frac{43}{4}$
(g) $\frac{56}{5}$ (h) $\frac{107}{10}$ (i) $\frac{36}{32}$ (j) $\frac{27}{4}$ (k) $\frac{135}{12}$

6 Change each of these mixed numbers to improper fractions:
(a) $2\frac{1}{4}$ (b) $3\frac{1}{2}$ (c) $4\frac{2}{5}$ (d) $5\frac{3}{8}$ (e) $8\frac{1}{5}$ (f) $10\frac{2}{3}$
(g) $3\frac{1}{4}$ (h) $5\frac{7}{8}$ (i) $4\frac{3}{5}$ (j) $12\frac{2}{3}$ (k) $25\frac{3}{4}$

7 Wasim bought a shirt and a tie. The cost of the shirt was £18. The total cost of the shirt and the tie was £22.
(a) Work out the cost of the tie as a fraction of:
  (i) the cost of the shirt
  (ii) the total cost of the shirt and tie
(b) Work out the cost of the shirt as an improper fraction of the cost of the tie.
(c) Change your answer to (b) into a mixed number.

Exercise 4.2   Links: *(4D–G)* 4D–G

1 Copy this shape. Find two equivalent fractions to describe how much of the shape has been shaded.

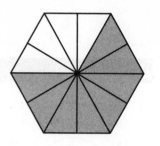

2 Simplify each of these fractions by finding common factors:
   (a) $\frac{6}{8}$ **(b)** $\frac{8}{10}$ **(c)** $\frac{12}{24}$ **(d)** $\frac{24}{36}$ **(e)** $\frac{75}{100}$ **(f)** $\frac{54}{90}$

3 Last season Lucea Hockey club played 40 matches. They won 32, lost 6 and drew 2. Writing each fraction in its simplest form, find the fraction of the 40 matches that Lucea Hockey club:
   (a) won
   (b) lost
   (c) drew

4 Find:
   (a) $\frac{1}{2}$ of 80   **(b)** $\frac{2}{3}$ of 60   **(c)** $\frac{4}{5}$ of 100
   (d) $\frac{3}{7}$ of 140   **(e)** $\frac{3}{4}$ of £28.40

5 A company makes $\frac{3}{8}$ of its 240 employees redundant.
   How many employees are made redundant?

6 Joan's gross pay is £1200 per month. Her stoppages are $\frac{2}{5}$ of her gross pay. Work out Joan's stoppages per month.

7 The cost of a new motor cycle is £3600. Andrew paid a deposit which is $\frac{5}{8}$ of the cost. Work out the deposit paid by Andrew.

8 Jaqui travelled 48 miles to see her sister. She travelled the first 18 miles by car and the remainder by rail. Work out, in their simplest form:
   (a) the fraction of the journey Jaqui travelled by car
   (b) the fraction $\dfrac{\text{distance travelled by car}}{\text{distance travelled by rail}}$

9 Make these fractions equivalent:

(a) $\frac{3}{4} = \frac{9}{}$ (b) $\frac{16}{20} = \frac{}{5}$ (c) $\frac{1}{3} = \frac{}{24}$ (d) $\frac{35}{100} = \frac{7}{}$

10 In each case write down three fractions equivalent to:

(a) $\frac{4}{5}$ (b) $\frac{2}{7}$ (c) $\frac{18}{24}$ (d) $\frac{27}{36}$ (e) $\frac{30}{100}$

11 Which is the larger:

(a) $\frac{2}{3}$ or $\frac{3}{5}$ (b) $\frac{3}{4}$ or $\frac{4}{5}$ (c) $\frac{7}{10}$ or $\frac{17}{20}$

12 Put these fractions in order, starting with the smallest:

$\frac{3}{10}$ $\frac{1}{3}$ $\frac{2}{7}$ $\frac{4}{15}$ $\frac{29}{100}$

Exercise 4.3   Links: *(4H–K)* 4H–K

1 Work out:

(a) $\frac{3}{7} + \frac{2}{7}$ (b) $\frac{4}{13} + \frac{8}{13}$ (c) $\frac{1}{8} + \frac{5}{8}$ (d) $\frac{2}{5} + \frac{3}{5}$

(e) $\frac{3}{8} + \frac{1}{2}$ (f) $\frac{1}{3} + \frac{1}{6}$ (g) $\frac{5}{12} + \frac{3}{4}$ (h) $\frac{3}{4} + \frac{9}{20}$

(i) $\frac{3}{5} + \frac{3}{10}$ (j) $\frac{2}{15} + \frac{1}{30}$ (k) $\frac{5}{12} + \frac{1}{3}$ (l) $\frac{2}{5} + \frac{7}{20}$

2 Work out:

(a) $\frac{1}{6} + \frac{3}{8}$ (b) $\frac{3}{5} + \frac{3}{8}$ (c) $\frac{2}{9} + \frac{1}{12}$ (d) $\frac{3}{10} + \frac{4}{15}$

(e) $\frac{1}{3} + \frac{1}{5}$ (f) $\frac{1}{7} + \frac{1}{6}$ (g) $\frac{2}{5} + \frac{3}{7}$ (h) $\frac{3}{5} + \frac{1}{8}$

(i) $\frac{4}{5} + \frac{1}{3}$ (j) $\frac{7}{12} + \frac{4}{9}$ (k) $3\frac{1}{2} + 2\frac{1}{4}$ (l) $4\frac{1}{3} + 3\frac{1}{4}$

3 George won some money in the lottery.
He gave $\frac{2}{5}$ of this money to his wife.
He gave $\frac{1}{3}$ of this money to his daughter.
What fraction of the money did his
wife and daughter receive altogether?

4 Work out:

(a) $\frac{7}{8} - \frac{1}{8}$ (b) $\frac{5}{8} - \frac{1}{2}$ (c) $\frac{1}{3} - \frac{1}{5}$ (d) $\frac{1}{2} - \frac{1}{7}$

(e) $\frac{3}{8} - \frac{1}{3}$ (f) $\frac{4}{5} - \frac{1}{7}$ (g) $\frac{1}{3} - \frac{1}{8}$ (h) $\frac{2}{3} - \frac{2}{5}$

(i) $\frac{5}{7} - \frac{3}{10}$ (j) $2\frac{1}{2} - \frac{1}{4}$ (k) $3\frac{1}{4} - \frac{1}{2}$ (l) $5 - 1\frac{3}{4}$

(m) $4\frac{1}{4} - 1\frac{3}{8}$ (n) $3\frac{1}{3} - 1\frac{1}{4}$ (o) $5\frac{2}{3} - 3\frac{7}{8}$ (p) $4 - 2\frac{3}{8}$

(q) $4\frac{2}{3} - 1\frac{1}{2}$ (r) $17\frac{1}{4} - 3\frac{2}{5}$

5 The farmer had $12\frac{1}{2}$ acres of land.
He sold $10\frac{5}{8}$ acres to a builder.
How much land did the farmer
have left after the sale?

6 Work out

    (a) $\frac{1}{2} \times \frac{1}{4}$ **(b)** $\frac{5}{8} \times \frac{1}{4}$ **(c)** $\frac{2}{3} \times \frac{4}{5}$

    (d) $\frac{3}{5} \times \frac{7}{8}$ **(e)** $\frac{2}{5} \times \frac{3}{4}$ **(f)** $\frac{3}{7} \times \frac{7}{12}$

    (g) $\frac{1}{4} \times 20$ **(h)** $\frac{2}{5} \times 8$ **(i)** $\frac{3}{14} \times 280$

    (j) $57 \times \frac{2}{3}$ **(k)** $24 \times \frac{3}{8}$ **(l)** $15 \times \frac{7}{8}$

    (m) $2\frac{1}{4} \times 3\frac{1}{2}$ **(n)** $4\frac{3}{4} \times 5\frac{1}{2}$ **(o)** $1\frac{1}{3} \times \frac{1}{4}$

    (p) $3\frac{2}{5} \times 4\frac{1}{7}$ **(q)** $6\frac{1}{2} \times 4\frac{1}{4}$ **(r)** $2\frac{1}{5} \times 1\frac{7}{8}$

7 A rectangular carpet measures $3\frac{1}{4}$ metres by $4\frac{1}{8}$ metres. Work
out the area of the carpet.

8 The capacity of a large jug is $6\frac{1}{3}$ litres.
Work out the total capacity of $3\frac{1}{2}$ of these jugs.

9 Work out:

    (a) $\frac{1}{2} \div \frac{1}{4}$ **(b)** $\frac{7}{8} \div \frac{1}{4}$ **(c)** $2\frac{3}{4} \div 1\frac{1}{2}$

    (d) $\frac{7}{8} \div 4$ **(e)** $\frac{2}{7} \div 2$ **(f)** $\frac{5}{7} \div 3$

    (g) $12 \div \frac{1}{4}$ **(h)** $7 \div \frac{1}{3}$ **(i)** $\frac{3}{7} \div \frac{1}{4}$

    (j) $\frac{4}{5} \div \frac{5}{9}$ **(k)** $2\frac{1}{2} \div 3\frac{1}{4}$ **(l)** $4\frac{2}{7} \div 3\frac{5}{8}$

5 Two-dimensional shapes

1 Complete the sentences using numbers, and words chosen from:

     equal   parallel   opposite   sides   angles

   (a) A triangle has 3 _____ and 3 _____ .

   (b) A quadrilateral has _____ sides.

   (c) _____ sides in a parallelogram are _____ and _____ .

   (d) Pairs of adjacent _____ in a kite are _____ .

   (e) The _____ of a square are equal.

   (f) An isosceles triangle has 2 _____ _____ and 2 _____ _____ .

   (g) The _____ of a pentagon always add up to _____ degrees.

   (h) All the _____ of a rhombus are _____ . The
       opposite _____ are _____ and _____ .

2 On squared paper draw:

   (a) 3 different triangles which have a base of 5 cm and a height
       of 3 cm

   (b) a trapezium with parallel sides 6 cm and 4 cm and with two
       angles as right angles

   (c) three different isosceles triangles with height 2 cm

   (d) a parallelogram with shortest sides 3 cm

3 Answer the questions about the 2-D shapes in the picture.

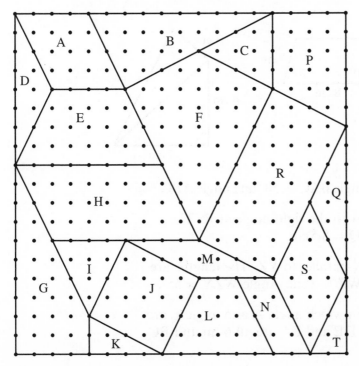

(a) List the shapes which are parallelograms
(b) What mathematical name describes the shape R?
(c) What mathematical name describes the shape J?
(d) Which shape is a rhombus?
(e) List the shapes which are trapeziums.
(f) Which shapes are right-angled triangles?
(g) Which shape is the kite?

4 Investigate the number of different quadrilaterals you can make using 2, 3 or 4 identical right-angled triangles. In each case, give their mathematical name, e.g.

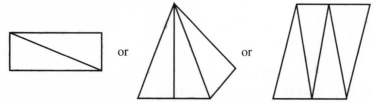

Exercise 5.2 Links: (5C) 5C

1 Construct, using ruler and compasses:
(a) a triangle with sides 8 cm, 5 cm and 6 cm
(b) a triangle with sides 7 cm, 6 cm and 6 cm
(c) a quadrilateral with sides 6 cm, 8 cm, 13 cm and 13 cm where one of the diagonals is 10 cm long

Using ruler, compasses and protractor make accurate drawings of the shapes in questions **2** to **6**. (Hint: first make a sketch so that you know what it looks like.)

2 (a) (b) (c)

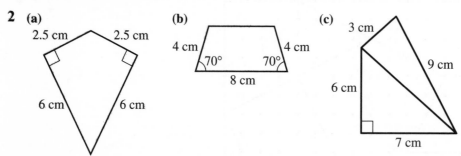

3 A triangle where AB = 5 cm, angle BAC = 50° and angle ABC = 65°.

4 A right-angled triangle PQR with the right angle at R. The lengths are QR = 6 cm and PQ = 8.5 cm.

5 A trapezium WXYZ with WX parallel to YZ. The lengths are WX = 5 cm, ZY = 7 cm and WZ = 3 cm. Angle WZY = 55°.

6 A convex pentagon ABCDE in which angle A and angle E are right angles. The lengths AB, DE and EA are all 6 cm and BC and CD are both 4 cm.

Exercise 5.3   Links: (5D) 5D, E

Write down the letters of the shapes that are congruent.

1

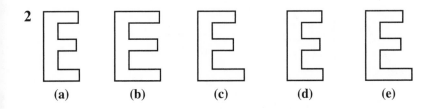

2

3

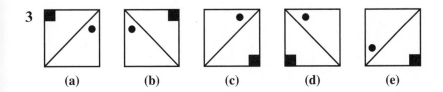

4

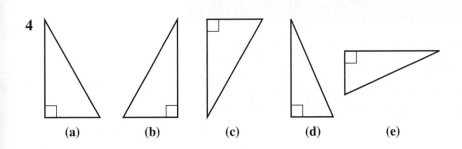

5 List the pairs of congruent shapes in the diagram.

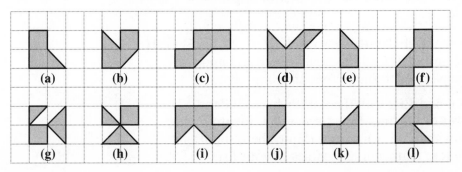

6 In each part, pick the two triangles which are congruent and give the reason.

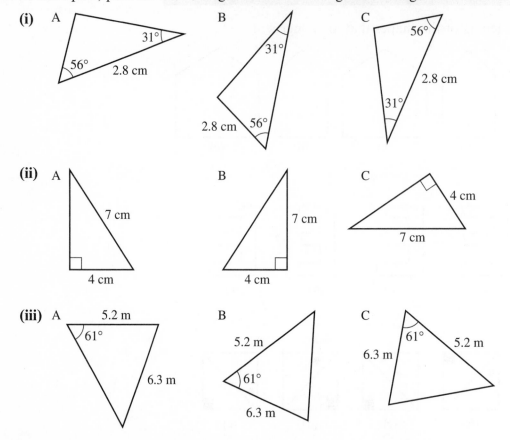

(i)   A   B   C

(ii) A   B   C

(iii) A   B   C

1 Work out the exterior angle of a regular hexagon. Write down the interior angle of a regular hexagon.

2 A twenty-sided shape is called an icosagon. Work out the exterior angle of a regular icosagon. Write down the sum of the interior angles of an icosagon.

3 A regular shape has an exterior angle of 20°. Work out how many sides it has. What is the size of an interior angle?

4 A nonagon has nine sides. Work out the interior angle for a regular nonagon.

5 Draw a regular octagon. Label the vertices A, B, C, D, E, F, G, H. Join AC, AD, CH and DH. Call the point where AD and CH cross X. Name the pairs of triangles that are congruent.

   Now join DG, EG, and EH. Call the point where DG and EH cross Y. Make a list of the lines that are parallel. Identify and name quadrilaterals that are congruent. (Hint: there should be 4 shapes with different mathematical names in your list.)

6 The shape shown can be used to tessellate the inside of a rectangle which is 6 squares by 9. Show on squared paper how this can be done.   [E]

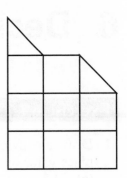

7 Show how these quadrilaterals tessellate.

(a)   (b)

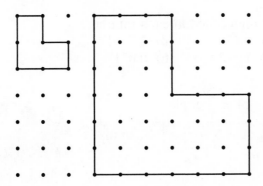

8 Tessellate the small 'L' shape into the enlarged 'L' shape.

6 Decimals

1 Draw a copy of the place value diagram. On your copy of the
   place value diagram write in these numbers

   (a) 34.7   (b) 3.47   (c) 0.347   (d) 243.75
   (e) 0.072   (f) 132.89   (g) 10.003   (h) 0.04

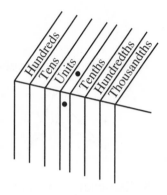

2 What is the place value of the digit underlined in each number?

   (a) 4<u>7</u>.9   (b) 4.<u>7</u>9   (c) 0.4<u>7</u>9   (d) 2<u>5</u>3.81
   (e) 18.1<u>0</u>6   (f) 24.03<u>8</u>   (g) 110.3<u>2</u>   (h) 0.07<u>1</u>

3 The table gives the heights in metres of six people.

Steven	Gemma	Fatima	Paul	Yvonne	Yasmin
1.83	1.57	1.63	1.85	1.68	1.75

   Write the list of names in increasing order of height, starting
   with the shortest.

4 Re-arrange these decimals in order of size, starting with the
   smallest.

   (a) 4.73, 7.43, 3.47, 4.07, 7.04
   (b) 15.03, 1.503, 5.103, 3.105, 1.035, 31.05
   (c) 0.0011, 10.10, 0.1001, 0.1100
   (d) 1.061, 1.106, 1.601, 1.016
   (e) 0.07, 0.70, 0.07, 0.007, 7.00
   (f) 5.16, 5.016, 5.061, 5.61

5 Re-arrange these decimals in order of size. Start with the largest.

   (a) 4.4, 4.12, 4.75, 7.45, 5.74
   (b) 0.011, 0.0011, 0.101, 0.110
   (c) 0.52, 0.075, 0.507, 0.4, 0.08
   (d) 0.07, 0.14, 0.008, 0.205, 0.025
   (e) 3.09, 2.08, 3.2, 2.3, 2.267
   (f) 2.222, 22.22, 222.2, 2222, 2.22222

6 Eight people took part in a 400 metres race. Their names and
   times, in seconds, for the race are given below:

Smith	104.82	Thomas	102.37
Warne	103.41	Kelly	103.47
Khan	102.15	McGowan	104.72
Young	105.28	Priest	103.94

   The winner is the person with the lowest time.
   (a) Who won the race?   **(b)** Who came last?

7 Round:
   (a) 18.6 correct to the nearest whole number
   (b) 102.48 correct to the nearest whole number
   (c) 3.53 correct to one place of decimals
   (d) 14.365 correct to one place of decimals
   (e) 25.057 correct to two places of decimals
   (f) 0.053 correct to one place of decimals

8 The winning time for a 200 metres race was 23.468 seconds.
   Round this number correct to the second decimal place.

9 The engine capacity, in litres, of Maureen's car is 1.398. Round
   this number correct to:
   (a) the nearest whole number of litres
   (b) one place of decimals

Exercise 6.2   **Links: (*6C–F*) 6D, E, F**

1 Work these out **without a calculator**, showing all your working.
   (a) 2.7 + 5.6   **(b)** 0.75 + 4   **(c)** 32.6 + 53.2
   (d) 324.8 + 2.25   **(e)** 349 + 34.9   **(f)** 0.017 + 1.235
   (g) 62 + 6.2   **(h)** 143.5 + 3.5   **(i)** 5.67 + 0.033

2 Work these out **without a calculator**, showing all your working.
   (a) 7.4 + 17.6 + 27.3   **(b)** 23.07 + 8.6 + 7.44
   (c) 0.723 + 1.1 + 12.453   **(d)** 3.8 + 3.8 + 3.8
   (e) 4.07 + 32 + 0.0457   **(f)** 214.073 + 92 + 5.91

3 Work these out **without a calculator**, showing all your working.
   (a) 28.8 − 17.3   **(b)** 6.72 − 2.6   **(c)** 32.8 − 8.2
   (d) 100 − 8.5   **(e)** 0.68 − 0.52   **(f)** 15 − 2.85
   (g) 1000 − 986.4   **(h)** 0.17 − 0.073   **(i)** 305.07 − 56.28

Exercise 6.3   Links: (*6G, H*) 6G, H, I

1 Work these out **without a calculator**, showing all your working.
   (a) The total cost of 5 CD's at £2.99 each
   (b) The total cost of 6 cans of drink at £0.45 each
   (c) The total cost of 12 litres of petrol at £0.74 per litre
   (d) The total cost of 2.5 kilos of apples at £1.18 per kilo

2 Work these out **without a calculator**, showing all your working.
   (a) 8.7×4 **(b)** 0.87×4 **(c)** 0.87×40
   (d) 1.3×10 **(e)** 23.5×12 **(f)** 32.5×22
   (g) 71.3×10 **(h)** 2.1×5 **(i)** 0.6×0.8
   (j) 0.35×0.6 **(k)** 1.7×0.3 **(l)** 0.1×0.1
   (m) 0.04×0.2 **(n)** 0.03×0.08 **(o)** 3.14×0.6
   (p) 0.4×0.4

3 Work these out **without a calculator**, showing all your working.
   (a) $81.6 \div 3$ **(b)** $27 \div 10$ **(c)** $47.5 \div 5$
   (d) $24.68 \div 4$ **(e)** $7.8 \div 20$ **(f)** $54.3 \div 6$
   (g) $37.8 \div 10$ **(h)** $0.46 \div 10$ **(i)** $57 \div 100$
   (j) $3 \div 100$ **(k)** $0.38 \div 100$ **(l)** $75 \div 1000$

4 Eight people share £194.80 equally.
   How much will each person receive?

5 How many 4 gallon tanks will be needed
   to hold 78.4 gallons of oil?

6 Work out **without a calculator**, showing all your working.
   (a) $4 \div 0.2$ **(b)** $8.64 \div 2.4$ **(c)** $19.2 \div 12.8$
   (d) $0.258 \div 0.3$ **(e)** $1.7 \div 0.5$ **(f)** $58.08 \div 12.1$
   (g) $0.003 \div 0.15$ **(h)** $3.2 \div 0.8$ **(i)** $360.4 \div 6.8$
   (j) $0.035 \div 0.01$ **(k)** $10.1 \div 0.4$ **(l)** $2.624 \div 6.4$

Exercise 6.4   Links: (*6I*) 6J

1 Change these fractions to decimals. Show your working.
   (a) $\frac{1}{4}$ **(b)** $\frac{2}{5}$ **(c)** $\frac{9}{10}$ **(d)** $\frac{7}{100}$ **(e)** $\frac{3}{20}$
   (f) $\frac{3}{8}$ **(g)** $\frac{1}{3}$ **(h)** $\frac{2}{7}$ **(i)** $\frac{37}{50}$ **(j)** $\frac{54}{90}$
   (k) $\frac{17}{200}$ **(l)** $\frac{31}{40}$ **(m)** $\frac{18}{35}$ **(n)** $\frac{637}{1000}$ **(o)** $\frac{13}{2000}$

2 Change these decimals to fractions.
   (a) 0.7   **(b)** 0.41   **(c)** 0.253   **(d)** 0.8
   (e) 0.35   **(f)** 0.173   **(g)** 0.99   **(h)** 0.09
   (i) 0.011   **(j)** 0.0041   **(k)** 0.0909   **(l)** 0.0035

3 Write the numbers in the cloud in order, starting with the smallest.

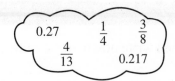

4 Write the numbers in the cloud in order, starting with the smallest.

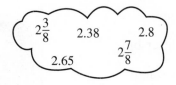

5 (a) Change $4\frac{3}{8}$ to decimals.
   (b) Work out the exact value of $4\frac{3}{8} \times 15.3$.
   (c) Round your answer to part **(b)**:
      (i) correct to the nearest whole number
      (ii) correct to one place of decimals
      (iii) correct to two places of decimals
      (iv) correct to three places of decimals

7 Measure 1

Exercise 7.1   **Links: (7A) 7A**

Look at this picture, then write down an estimate for each of the following.

1 The height of the bus

2 The width of the bus

3 The height of the litter bin

4 The height of the woman

5 The height of the lamppost

6 The diameter of the bicycle wheel

7 The length of the car

8 The height of the car

9 The diameter of the wheel

10 The height of the shop window from the ground

11 The width of the shop window

12 The width of the car

Exercise 7.2   **Links: (7B) 7B**

Write down an estimate of the amount of liquid in each of these containers. Give your answers in metric and Imperial units.

1 The coffee cup

2 The coffee maker

3 The mug of tea

4 The glass of milk with a straw in it

5 The wine glass

6 The bottle of champagne

7 The jug when it is full

8 The glass next to the jug

Exercise 7.3   Links: (*7C*) 7C

Write down an estimate of the weight of the following fruits. Give
your answer in metric and Imperial units.

1 The 4 bananas

2 The 3 kiwi fruit

3 The raspberries

4 The 5 pears

5 The pineapple

6 The bowl of blackberries

7 A large bag of potatoes

8 A cabbage

9 A bag of onions

10 A box of apples

11 A basket of potatoes

12 A melon

Exercise 7.4   Links: (*7D*) 7D

For each of these statements say whether the measurements are
sensible or not. If the statement is not sensible then give a reasonable
estimate for the measurement.

1 My mother is 1.60 m tall.

2 My 15-year-old brother is 2 cm tall.

3 The front door of my house is 20 m high.

4 My bedroom measures 30 m by 20 m.

5 An apple weighs 2 kg.

6 A can of cola holds about 400 ml of liquid.

7 This page is about 27 mm long.

8 A loaf of bread weighs about 800 g.

9 A box of chocolates weighs about 500 kg.

10 This book weighs about 10 kg.

11 The petrol tank in a small car holds about 50 litres.

12 A large bottle of cola holds about 3 ml of liquid.

13 A tea cup, when full, holds about 250 ml.

14 My car is capable of travelling at 500 km per hour.

15 A high-speed train can travel at 100 miles an hour.

Exercise 7.5   **Links: (*7E*) 7E**

Copy and complete this table with appropriate units for each
measurement. Give both metric and Imperial units of measurement.

	Metric	Imperial
1 Your height		
2 The length of your desk		
3 The weight of a packet of chocolates		
4 The weight of a large bag of potatoes		
5 Your weight		
6 The amount of water in a water barrel		
7 The amount of liquid in a petrol tanker		
8 The amount of wine in a wine bottle		
9 The time it takes to walk one mile		
10 The time it takes to run 100 metres		
11 The time it takes to drive 500 miles		
12 The thickness of this book		
13 The weight of a tube of sweets		
14 The time it takes to travel from Earth to Jupiter		
15 The diameter of a football		

Exercise 7.6   **Links: (*7F, G, H*) 7F, G, H**

Write down the time shown on these clocks.

1    2    3    4

5 Bill arrives at the bus station and looks at
   his watch. The next bus is due to arrive at 5:30 pm.
   How long does Bill have to wait until the bus
   should arrive?

6 Draw 4 clock faces and mark these times on them:
   (a) quarter past 3   **(b)** quarter to 9
   (c) twenty past 4   **(d)** ten to 5

7 Draw 4 digital watches and mark these times on them:
    (a) quarter to 6   **(b)** half past 2
    (c) twenty five to 7   **(d)** five past 4

8 Seamus arrives at the train station
and looks at the time shown on the
station clock. His train is due to arrive
at 10:35 am.

    (a) How long does he have to wait
until the train should arrive?

The train arrives 6 minutes late.
    (b) How long did Seamus have
to wait for his train?

9 Change these times from 12-hour clock times (am or pm) to
24-hour clock times.
    (a) 9:00 am   **(b)** 9:00 pm
    (c) 6:30 am   **(d)** 6:45 pm
    (e) 1 am   **(f)** 2 pm
    (g) a quarter to eight in the morning
    (h) a quarter past seven in the evening

10 Change these times from 24-hour clock times to 12-hour clock
times (am or pm).
    (a) 07:00   **(b)** 17:00
    (c) 15:30   **(d)** 08:50
    (e) 18:50   **(f)** 07:30
    (g) 00:10   **(h)** 23:45

Exercise 7.7   Links: (*7I–N*) 7I–N

1 Write down the readings on these scales.

    (a)

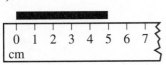

    (b)

    (c)

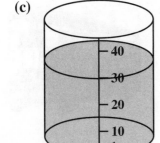

    (d)

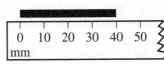

2 Write down the readings on these scales.

(a)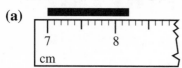

(b)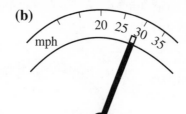

(c)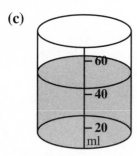

(d)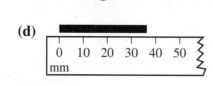

3 Write down the cost of posting these letters by **(i)** first class **(ii)** second class.

(a)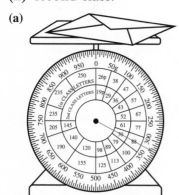

(b)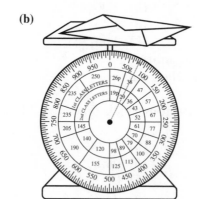

4 Measure and write down the lengths of these lines in centimetres.

(a) _____

(b) _____

(c) _____

(d) _____

(e) _____

5 Draw and label lines of length:
(a) 6 cm	**(b)** 5.5 cm	**(c)** 3.8 cm	**(d)** 10.3 cm
(e) 23 mm	**(f)** 55 mm	**(g)** 30 mm	**(h)** 112 mm

6 Write down the units you would use to measure:
(a) the length of the double decker bus
(b) the height of the double decker bus
(c) the weight of the double decker bus
(d) the capacity of the fuel tank of the double decker bus

7 Write down the units you would use to measure:
(a) the distance of your school from your home
(b) the time it takes you to travel from home to school

8 Collecting and recording data

Here are some questions that do not work properly.
Improve each one and say what was wrong with the original.

1 Terry wants to carry out a survey on the number of hours that
   people watch television. He asks:
   'Do you watch television?'

   a lot ☐   a little ☐   never ☐

2 Sybil is carrying out a survey on hotels.
   She asks, 'Was everything O.K.?'

3 Anthea was carrying out a survey on favourite breakfast cereals.
   She asks:
   'Do you like cereals?'

4 Teresa was carrying out a survey on the type of films people like
   watching. She asks:
   'What films do you like?'

   comedy ☐   sci. fi. ☐   adventure ☐

5 Michael wants to find out what type of music people like to
   listen to. He asks:
   'Do you like pop music or classical music?'

6 Cedric wants to know what type of weather people had on their
   holiday. He asks:
   'What was the weather like on your holiday?'

   good ☐   terrible ☐   OK ☐

7 David carries out a survey on which party people voted for in
   the general election. He asks:
   'You did vote for the Monster Raving Loony Party, didn't you?'

   Yes ☐   No ☐

8 Kate asks her friends about their mock exam results. She asks:
   'How well did you do in your mock exams?'

   badly ☐   well ☐   OK ☐

9 Draw up a questionnaire to find out what type of food people
   eat during the course of a school day.

BALLOT
Place your **X** in one box only
Labour ☐
Liberal Democrats ☐
Conservative ☐
Monster Raving Loony ☐

10 Design a questionnaire to find out what people are looking for when they choose a car.

Exercise 8.2   Links: *(8C, D, E)* 8C, D, E

1 In each of these cases choose which type of sampling technique you would use to collect your data.
  (a) Where people eat their lunch
     (i) by asking the first 10 people in the lunch queue
    (ii) by asking every third person in your tutor group
   (iii) by asking the first 10 people going home for lunch.
  (b) What you had for breakfast
     (i) by asking every third person in your tutor group
    (ii) by asking all those people who were late for school
   (iii) by asking all those people on the school bus.
  (c) How you travelled to school
     (i) by asking every third person in your tutor group
    (ii) by asking all those people who were late for school
   (iii) by asking all those people on the school bus.

2 A new supermarket is to be built. The company carry out a survey on where it should be built. There are three sites where it could be built. One site is in the middle of town near 3 other supermarkets, one site is on the edge of town with its own free car park and the other site is next to the football ground.
  (a) Suggest a suitable sample and where it should be collected from.
  (b) Design a suitable questionnaire that the company could use to collect their data.

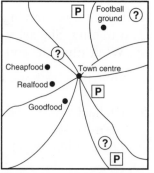

? Proposed new supermarket

3 Design a suitable data collection sheet to find out how long the members of your class spend on their homework each week. Use your data collection sheet to check how easy it is to use.

4 Design a suitable data collection sheet for finding out what type of vehicle passes the school gate at the end of the school day. Use your data collection sheet to check how easy it is to use.

5 Carry out the following experiments and record your results in a tally table. You should repeat each experiment 50 times.
  (a) Roll two dice, add the scores on each one and record the result.
  (b) Roll two dice, subtract the scores on each one and record the result.
  (c) Throw two coins in the air and record the number of times they come down with 2 heads or 2 tails or a head and a tail.

Exercise 8.3   Links: *(8F)* 8F, G

1 Here is some data about the students in Gareth's maths class

Name	Tutor Group	Test mark	Estimated GCSE grade	Number of detentions	Number of merits	Parents seen
Gareth	11 A	50%	E	0	5	Yes
Naomi	11 A	46%	E	2	2	Yes
Nilmini	11 C	67%	D	1	1	No
Moshe	11 D	72%	D	0	6	Yes
Eira	11 A	35%	F	4	2	No
Flora	11 C	42%	E	2	3	No
Iain	11 C	55%	E	4	0	Yes

Use the information to answer the following questions:
(a) Which students scored more marks in the test than Gareth?
(b) Which tutor group was only one person in?
(c) Write down the names, in the order of most merits received.
(d) How many students had their parents seen?
(e) Which student had the lowest estimated GCSE grade?
(f) Write down the names, in the order of best test mark.
(g) Which students had no detentions?

2 Here is some data about three of the second hand cars available from a local garage.

Make	Model	Year	Size of engine	Number of miles	Number of owners	Price
Volvo	850 SE	1997	2500	12 000	2	£15 500
Saab	9000i	1996	2000	25 000	3	£10 500
Honda	Civic	1996	2000	30 000	2	£8500
Rover	420i	1995	1600	45 000	1	£7995

(a) Which is the cheapest car?
(b) Which car has the biggest engine?
(c) Write down the cars in order of price.
(d) Which car is the oldest?
(e) Which car has had the most owners?

3

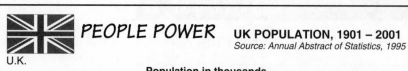

PEOPLE POWER   UK POPULATION, 1901 – 2001

Source: Annual Abstract of Statistics, 1995

U.K.

Population in thousands

	UNITED KINGDOM	ENGLAND	WALES	SCOTLAND	NORTHERN IRELAND
1901	38,237	30,515	2,013	4,472	1,237
1911	42,082	33,649	2,421	4,761	1,251
1921	44,027	35,231	2,656	4,882	1,258
1931	46,038	37,359	2,593	4,843	1,243
1951	50,225	41,159	2,599	5,096	1,371
1961	52,807	43,561	2,635	5,184	1,427
1971	55,928	46,412	2,740	5,236	1,540
1981	56,352	46,821	2,813	5,180	1,538
1993	58,191	48,533	2,906	5,120	1,632
2001	59,800	50,023	2,966	5,143	1,667

(a) What was the population of Wales in 1971?

(b) How much did the population of the United Kingdom rise between 1961 and 1971?

(c) In what year was the population of Scotland the highest?

(d) Between which years did the population of Northern Ireland fall and then rise again?

(e) Between which years did the population of England rise the least?

9 Algebra 2

1

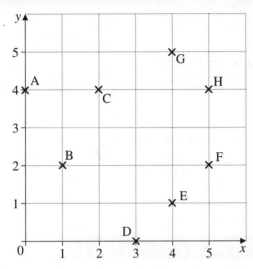

(a) Write down the letters on the grid with these coordinates:
   (i) (4, 1)   **(ii)** (2, 4)   **(iii)** (0, 4)   **(iv)** (4, 5)
   (v) (5, 4)   **(vi)** (3, 0)   **(vii)** (1, 2)   **(viii)** (5, 2)

(b) Write down the coordinates of these letters on the grid:
   (i) D   **(ii)** H   **(iii)** C   **(iv)** E

2 Write down the coordinates of all the points marked that make up this shape:

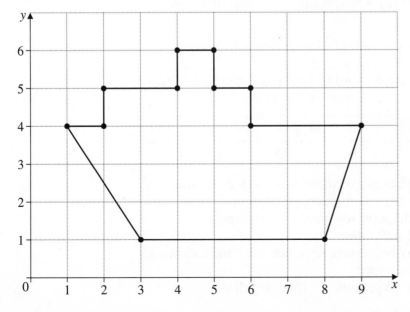

3   Draw a coordinate grid from 0 to 10 in both directions. On your
    grid plot the points and draw the following shapes by joining the
    points in order.
    (a) (1, 1), (1, 6), (2, 7), (7, 7), (8, 6), (8, 1), (1, 1)
    (b) (4, 1), (4, 3), (5, 3), (5, 1)
    (c) (4, 7), (4, 9), (6, 9), (6, 7)
    (d) (2, 4), (2, 6), (4, 6), (4, 4), (2, 4)
    (e) (5, 4), (5, 6), (7, 6), (7, 4), (5, 4)

4   Draw a coordinate grid from 0 to 10 in both directions. On your
    grid draw a design of your own. Label each point and write
    down its coordinate.

5   Work out the co-ordinates of the mid-point of the line segments:
    (a) A(0, 0)   B(6, 8)   (b) C(2, 5)   D(4, 8)
    (c) E(3, 7)   F(4, 11)   (d) G(1, 2)   H(9, 7)

Exercise 9.2   Links: (*9C, D*) 9C, D

1   The table shows the cost of Brussels sprouts per lb.
    (a) Draw a graph for this table.

Weight in lb	1	2	3	4	5
Cost in pence	40	80	120	160	200

    (b) Work out the cost of:
        (i) $2\frac{1}{2}$ lb sprouts   (ii) $4\frac{1}{2}$ lb sprouts
    (c) Extend the graph to work out the cost of:
        (i) 6 lb sprouts   (ii) 8 lb sprouts   (iii) 10 lb sprouts

2   The table shows the distance travelled by a four-wheel drive
    vehicle for every gallon of diesel it uses.
    Copy and complete the table showing how much diesel the four-
    wheel drive uses.

Distance travelled (miles)	25	50	75	100	125	150
Diesel used (gallons)	1	2	3			

    (a) Copy and complete the table to show how much diesel was used.
    (b) Draw a graph from the information given in your table.
    (c) Work out how much diesel was used in travelling:
        (i) 40 miles   (ii) 110 miles
    (d) Work out how many miles were travelled after the following
        amounts of diesel were used.
        (i) 8 gallons   (ii) 10 gallons   (iii) 20 gallons

3 (a) Draw a conversion graph from litres to pints.
Use the fact that: 0 litres = 0 pints and 50 litres = 90 pints.
On your graph use these scales for litres and pints:
1 cm = 5 litres and 1 cm = 5 pints.
Plot the points (0, 0) and (50, 90) and join them with a
straight line.

(b) Copy and complete this table using your conversion graph
to help you.

Litres	0			38		25	44	28	15			50
Pints	0	10	20		36					50	72	90

4 Copy this table which converts ounces to grams. Use the
information to draw a conversion graph from ounces to grams.
Use your graph to help fill in the missing values.

Ounces (oz)	1	2			16	10	24	19
Grams (g)	28		113	340			680	

5 Use your graph in question 4 to find the number of grammes for
8 ounces.

6 (a) Draw a graph using the information below to show the
speed of a car and distance travelled in a drag race.

Speed in km/hr	0	25	60	150	190	200
Distance in metres	0	80	250	600	800	1000

(b) Use your graph to work out the speed when the distance
travelled is 500 m.
(c) Use your graph to work out the distance travelled when the
speed is 100 km/hr.

Exercise 9.3   **Links: (*9E, F*) 9E, F**

1 June went shopping by van. She drove 15 miles to the shops in
20 minutes. She stayed at the shops for 45 minutes and then
started to drive home. After 10 minutes, when she had driven
5 miles, she stopped for petrol for 5 minutes. It then took her a
further 15 minutes to get home.
Draw a distance time graph for June's journey.

2 Mustaq walks to the cafe, meets some friends, then walks home again.

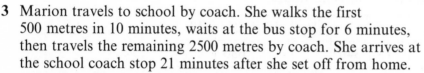

  (a) How many minutes did it take Mustaq to walk to the cafe?
  (b) How far away was the cafe?
  (c) How many minutes did Mustaq spend at the cafe?
  (d) How long did it take Mustaq to walk home?
  (e) Work out the speed, in metres per minute, for Mustaq to walk to the cafe. Also give your answer in kilometres per hour.
  (f) Work out the speed, in metres per minute, for Mustaq to walk back from the cafe. Give your answer in kilometres per hour as well.

3 Marion travels to school by coach. She walks the first 500 metres in 10 minutes, waits at the bus stop for 6 minutes, then travels the remaining 2500 metres by coach. She arrives at the school coach stop 21 minutes after she set off from home.
  (a) Draw a distance-time graph of her journey.
  (b) Work out the speed of the coach, first in metres per minute, then in kilometres per hour.

4 Write down all the coordinates of the points labelled A to L on the coordinate grid.

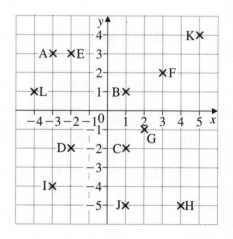

5 Draw a coordinate grid with the horizontal axis (*x*-axis) marked from −10 to +10 and the vertical axis (*y*-axis) marked from −5 to +5.
Plot the following points and join them in the order of this list:

   (−10, 0), (−10, 2), (−5, 4), (−4, 5), (−2, 1), (2, 3), (4, 2),
   (9, 3), (10, 2), (10, −2), (9, −3), (4, −2), (2, −4),
   (−2, −1), (−4, −5), (−5, −4), (−10, −2), (−10, 0)

Exercise 9.4   Links: (*9G, H, I*) 9G–J

1 (a) Copy and complete the tables of values for the graphs below.

(i) $y = 4x + 1$

x	−2	−1	0	1	2
y					

(ii) $y = 3 - 2x$

x	−2	−1	0	1	2
y					

(b) Draw the two graphs on the same grid and write down the coordinates of the point where they cross.

2 Write down the equations of the lines marked **(a)** to **(e)** in this diagram.

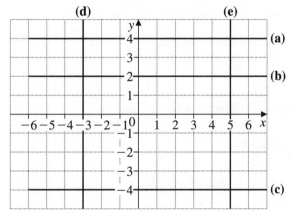

3 Draw a coordinate grid with *x*- and *y*-axes labelled from −5 to +5. On the grid draw and label the graphs of:
(a) $x = 1$ **(b)** $x = 3$ **(c)** $x = -5$ **(d)** $x = -2$
(e) $y = 3$ **(f)** $y = -1$ **(g)** $y = 1$ **(h)** $y = -3$

4 Draw a coordinate grid with axes labelled from −5 to +5. On the grid draw and label the graphs of:
(a) $y = -3$
(b) $x = +2$
(c) Write down the coordinates of the point where the two lines cross.

5 On a coordinate grid with the *x*-axis labelled from −5 to +5 and the *y*-axis labelled −5 to +5 draw the following graphs.
(a) $y = x$ **(b)** $y = x + 3$ **(c)** $y = x - 3$
(d) $y = x + 5$ **(e)** $y = x - 5$

6 On a coordinate grid with the *x*-axis labelled −4 to +4 and the *y*-axis labelled −12 to +12 draw the following graphs.
(a) $y = 3x$ **(b)** $y = 3x + 8$
(c) $y = -3x + 6$ **(d)** $y = -2x - 4$

7 (a) Copy and complete the tables of values for the graphs below.

$y = 2x - 3$ $y = -x + 2$

x	-2	-1	0	1	2	3	4
y							

x	-2	-1	0	1	2	3	4
y							

(b) Draw the graphs on graph paper.
(c) Write down the coordinates of the point when they cross.

8 (a) Copy and complete the table of values of the graph
$y = x^2 + 3$.

x	-3	-2	-1	0	1	2	3
$y = x^2 + 3$	12						12

(b) Draw a coordinate grid with values of x from -3 to $+3$ and values of y from 0 to 12. Plot the points from the table on the grid and join them with a smooth curve.

9 Draw a coordinate grid with values of x from -4 to $+4$ and values of y from -6 to $+12$.
  (a) Make a table of values for equations
      (i) $y = x^2 - 5$ **(ii)** $y = 2x + 3$.
  (b) Plot the coordinates for each of the tables of values and draw each graph.

10 State whether each shape is 1-D, 2-D or 3-D.
  (a) octagon   **(b)** rectangle   **(c)** cuboid

11 Write down all 3-dimensional co-ordinates from this list;
    (123), (3, 5, 7), (9, 3), (7), (25, 6), (2, 3, 4), (9, 2, 3), (3)

10 Sorting and presenting data

1 Yvonne and Gulzar conducted a survey into the colours of cars.
   They recorded the colour of 50 cars which passed the school
   gates one morning. The results of the survey are given below.

blue	white	red	red	grey
green	white	black	yellow	red
blue	grey	red	green	red
white	red	blue	yellow	blue
red	white	grey	red	red
black	red	red	green	blue
red	blue	brown	red	grey
silver	red	silver	red	white
black	grey	grey	white	black
yellow	green	red	blue	red

   (a) Draw up a tally chart and frequency table for these colours.
   (b) What was the most popular colour for a car?
   (c) Draw a bar chart to show the results of the survey.

2 As part of a coursework, Kevin asked 40 people which day of the
   week they were most likely to go to the shops to do their main
   shopping for the week. The results of this survey are shown below.

Sat	Tues	Sat	Thurs	Fri	Thurs	Wed	Sat
Sun	Wed	Thurs	Fri	Tues	Sat	Mon	Thurs
Thurs	Sat	Fri	Thurs	Thurs	Sat	Thurs	Wed
Sat	Thurs	Thurs	Sat	Thurs	Thurs	Thurs	Fri
Fri	Sat	Thurs	Tues	Sun	Fri	Sat	Thurs

   For this information:
   (a) Draw up a tally chart and frequency table.
   (b) Draw a bar chart.
   (c) Which is
       (i) the most popular day for people to do their main shopping
       (ii) the least popular day for people to do their main shopping?

3 Fifty students sat an examination in English. The number of
   spelling mistakes they made are recorded in the table below.

Spelling mistakes

6	5	2	8	10	5	9	8	1	12
10	8	7	7	13	12	2	3	8	7
7	6	15	12	10	8	5	11	15	5
8	8	2	6	7	2	8	12	14	3
4	9	8	5	4	14	12	8	3	8

For this information:
(a) Draw up a tally and frequency chart.
(b) Draw a bar chart.
(c) Find the mode of the number of mistakes.
(d) Find the range of the number of mistakes.

4 There are 80 members of Wellshall Theatre Society. The ages, in
   years, of the members are given below.

Ages in years

27	19	32	46	8	17	33	42	64	71
14	34	33	52	30	41	38	32	24	61
35	36	29	42	57	34	36	19	18	45
36	22	28	42	50	17	73	7	12	62
42	55	26	32	33	39	27	31	28	44
33	46	23	17	48	41	37	32	28	66
78	32	45	33	20	18	44	53	38	39
34	31	42	23	16	14	58	37	30	22

(a) Draw up a frequency table using class intervals of 0 to 9, 10
    to 19, etc.
(b) Illustrate the data with a bar chart.
(c) Work out the range of the ages.
(d) Write down the modal class interval.

The society is to accept some new members.
(e) What is the minimum number of new members the society
    needs to accept to bring about a change in the modal class
    interval.
(f) Make a comment about the ages of these new members if
    this change occurs.

Exercise 10.2   Links: (*10C, D*) 10C, D

1 The average daily temperature in Leeds and Athens is recorded
   in the table below in °C.

	Oct	Nov	Dec	Jan	Feb	Mar	Apr
Leeds	14	10	8	7	9	10	15
Athens	25	22	14	13	15	17	20

(a) Draw a dual bar chart to illustrate this data.
(b) Write down three statements about the temperatures in
    Leeds and Athens.

2 The pictogram shows some information about the number of
people who played at the local squash club last week.

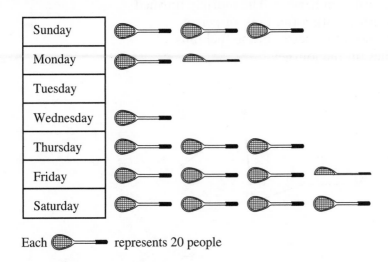

Each ⬤━━▬ represents 20 people

(a) How many played on
    (i) Sunday   **(ii)** Monday?

On Tuesday 50 people played squash at the club.
(b) Complete the pictogram.
(c) Work out the total number of people who played squash at
    the club during the week.
(d) Which was
    (i) the most popular day for people to play squash
    (ii) the least popular day for people to play squash.

It costs each person £1.50 to play squash at the club.
(e) What was the total amount people paid to play squash on
    (i) Friday   **(ii)** Tuesday?

3 A travel agent kept information about
the countries their customers visited
for their summer holiday last year.
This information was displayed in
the form of a pictogram
shown opposite.
(a) How many people visited
    (i) the United Kingdom
    (ii) Greece?

Last summer 300 customers visited
Turkey.
(b) Copy and complete the pictogram.
(c) Which was the most popular place
    that the customers visited last summer?

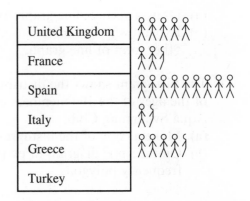

Each 🧍 represents 50 customers

Exercise 10.3   Links: (*10E, F*) 10E, F, G

1   At 8 am Jacqui started a journey in her car. The journey finished
    at 5 pm. She started the journey with a full tank of petrol.
    The line graph shows the amount of petrol in the fuel tank of
    Jacqui's car at various times on the journey.

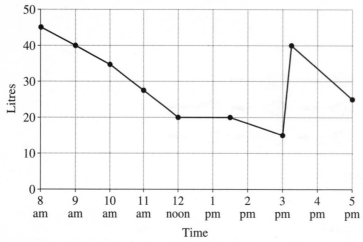

    (a) How much petrol does the fuel tank hold when full?
    (b) Between what times do you think Jacqui stopped for a lunch break?
    (c) Describe what you believe happened between 3:00 pm and 3:15 pm.
    (d) How much petrol was in the tank at
         (i) 10:00 am   (ii) 2:00 pm   (iii) the end of the journey?
    (e) At what time was the fuel tank half full?
    (f) How much petrol did the car use on the whole journey?

2   The table shows the number of students at Lucea High School
    who were late for the start of school one week.

Day	Mon	Tues	Wed	Thur	Fri
Number late	40	32	35	47	18

    (a) Draw a vertical line graph to display this information.
    (b) Explain why this information cannot be displayed using any
         other form of line graph.

3   The histogram shows the distribution
    of the ages of the 100 members of
    Aqua Swimming Club.
    (a) Make a copy of the histogram.
    (b) On the same diagram draw the
         frequency polygon.

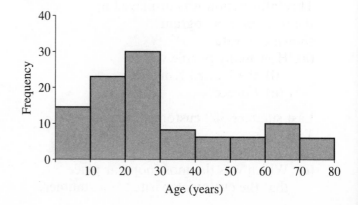

The distribution of the ages of the 100 members of Lucea Swimming Club is given below.

Age	10–19	20–29	30–39	40–49	50–59	60–69	70–79
Frequency	6	15	24	28	17	8	2

(c) For this second distribution, draw on the same axes
   (i) the histogram   (ii) the frequency polygon.
(d) Use the frequency polygons to compare the two distributions of ages, writing down three observations you have found.

4 Five thousand runners took part in a marathon race. The histogram shows part of the distribution of the times it took the runners to complete the race.

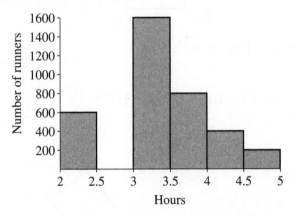

(a) How many runners completed the race in less than 2.5 hours?

1400 runners completed the race in more than 2.5 but less than 3 hours.
(b) Copy and complete the histogram.
(c) How many people completed the race in
   (i) less than 3 hours   (ii) more than 3.5 hours?

5 There are 120 people living in the small hamlet of Fellworth. The age ranges of these people are displayed in the table below.

Age (years)	0–9	10–19	20–29	30–39	40–49	50–59	60–69	70–79
Frequency	10	15	17	28	22	15	11	2

(a) Draw a histogram for this data.
(b) On your histogram, draw a frequency polygon.
(c) What is the modal age range for the people living in Fellworth?
(d) What percentage of the people living in Fellworth are aged between 20 and 39 years?

11 Three-dimensional shapes

1 ABCDEFGH is a cube resting on a level plane.
   (a) List the edges parallel to BC.
   (b) List the edges parallel to DH.
   (c) List the edges perpendicular to AD.
   (d) List the edges perpendicular to FG.
   (e) What face is parallel to ADCB?
   (f) What faces are perpendicular to CDHG?
   (g) Which faces are horizontal?
   (h) Which faces are vertical?

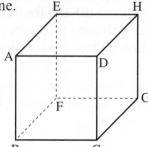

Remember:
Lines and faces are
perpendicular if the
angle between them is a
right angle.

2 ABCDEF is a triangular-based prism resting on a level plane.
   (a) Which faces are vertical?
   (b) Which faces are parallel?
   (c) List edges which are parallel to (i) EF   (ii) AB   (iii) DF
   (d) List edges which are horizontal.

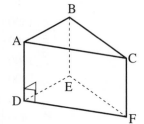

3 The boot stand PQRSTUVW is a prism with a trapezium
   cross section. It is resting on a level plane.
   (a) Which faces are vertical?
   (b) Which faces are perpendicular to TUVW?
   (c) Which edges are perpendicular to QR?
   (d) Which edges are parallel to PS?
   (e) Which edges are horizontal?

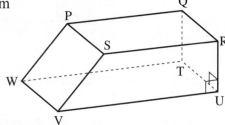

If possible, use triangular spotty paper.

1 The diagram shows the sketch of a 3-D letter H.

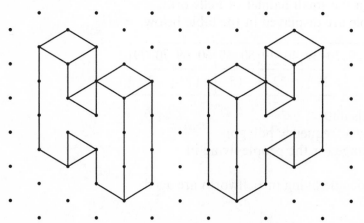

Make sketches for 3-D letters
(i) I   **(ii)** F   **(iii)** E   **(iv)** L   **(v)** T
You can also try letters A, B, C, O, P, U.

2 Sketch a hexagonal prism with its hexagonal faces horizontal. If you are using triangular spotty paper do one upright and one which leans to the side.

3 Name the shapes shown below.

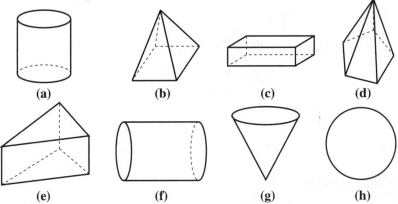

    (a)   **(b)**   **(c)**   **(d)**

    (e)   **(f)**   **(g)**   **(h)**

Exercise 11.3   Links: (*11F*) 11F

1 Draw sketches of the shapes that can be made from these nets.

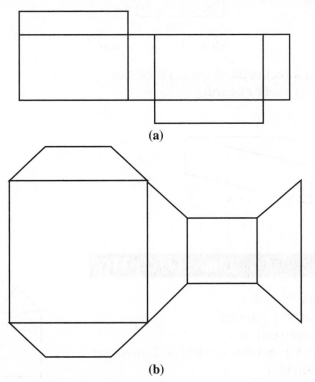

(a)

(b)

2 Which of the following are nets of a square-based pyramid?

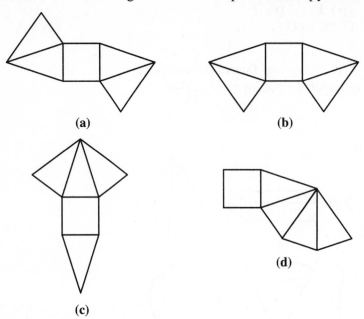

(a)   **(b)**

(c)   **(d)**

3 Draw accurate nets for the shapes below.

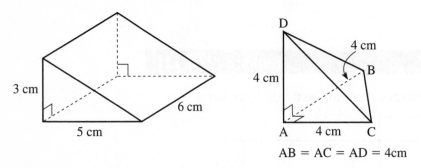

AB = AC = AD = 4cm

4 This cuboid is to be made from a rectangle of card. Draw the net that will fit on to the smallest sheet of card.

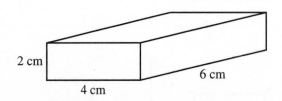

Exercise 11.4   Links: (*11G*) 11G

1 Copy the picture and on it mark
   (a) two lines that are horizontal and parallel
   (b) two lines that are parallel and vertical
   (c) two lines which are parallel but neither vertical or horizontal
   (d) two lines which are perpendicular

2 Give the full mathematical names of the following solids.

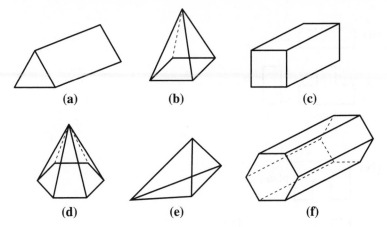

(a)   (b)   (c)

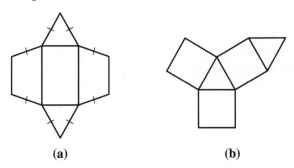

(d)   (e)   (f)

3 Sketch the shapes that these nets will make.

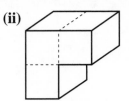

(a)   (b)

Exercise 11.5   Links: 11H

1 Sketch the plan and elevation of these solid shapes.

(i) 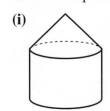   (ii)   (iii)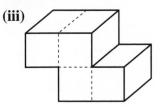

(iv) 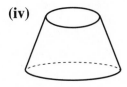   (v) 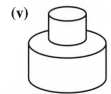   (vi)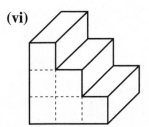

2 Sketch the solid shapes represented by these plans and elevations.

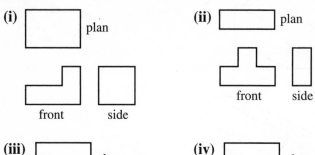

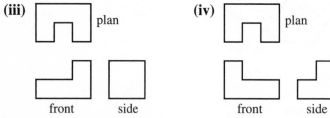

> *Note* :   There are no Practice exercises for
>   Unit 12: Using and applying mathematics.

13 Measure 2

1 Change these lengths to centimetres.
(a) 5 metres   (b) 11 metres   (c) 20 mm   (d) 300 mm
(e) 3.4 m   (f) 55 mm   (g) 5.7 m   (h) 254 mm

2 Change these lengths to millimetres.
(a) 2 cm   (b) 25 cm   (c) 3 m   (d) 4.55 m
(e) 0.5 cm   (f) 4.56 cm   (g) 0.5 m   (h) 0.03 m

3 Write these lengths in metres.
(a) 4 km   (b) 0.5 km   (c) 4500 cm   (d) 10 000 mm
(e) 5.43 km   (f) 55 km   (g) 0.03 km   (h) 300 cm

4 Change these weights to grams.
(a) 3 kg   (b) 50 kg   (c) 450 kg   (d) 2 tonnes
(e) 0.5 kg   (f) 0.35 kg   (g) 0.004 kg   (h) 12.5 kg

5 Change these weights to kilograms.
(a) 5000 g   (b) 50 000 g   (c) 5 tonnes   (d) 30 tonnes
(e) 450 g   (f) 0.6 tonnes (g) 250 g   (h) 0.003 tonnes

6 Write these weights in tonnes.
(a) 5000 kg   (b) 600 kg   (c) 75 kg   (d) 3500 kg
(e) 39 000 kg   (f) 75 500 kg   (g) 50 kg   (h) 55 kg

7 Change these lengths to kilometres.
(a) 4000 m   (b) 700 m   (c) 3500 m   (d) 60 000 m
(e) 500 m   (f) 235 m   (g) 3450 m   (h) 50 m

8 Write these volumes in millilitres.
(a) 4 *l*   (b) 3.5 *l*   (c) 60 *l*   (d) 0.5 *l*
(e) 4.85 *l*   (f) 0.06 *l*   (g) 0.003 *l* (h) 2.05 *l*

9 Change these volumes into litres.
(a) 5000 m*l*   (b) 4500 m*l*   (c) 600 m*l*   (d) 50 m*l*
(e) 10 m*l*   (f) 650 m*l*   (g) 25 m*l*   (h) 10 000 m*l*

10 How many 250 m*l* glasses can be filled from a 3-litre bottle of cola?

11 How many 2.5 centimetres are there in 150 metres?

Exercise 13.2   Links: (*13B, D*) 13B, D

1 Put these lines in order, smallest first.

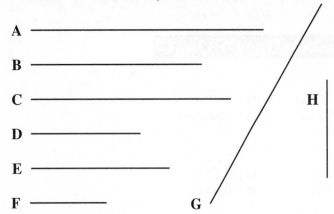

2 Write these lengths in order. Put the smallest first.
   (a) 5 m, 10 m, 6.5 m, 8 m, 9.5 m, 5.5 m
   (b) 400 cm, 5 m, 4.5 m, 475 cm, 550 cm, 5000 mm
   (c) 2 cm, 25 mm, 0.1 m, 30 mm, 2.8 cm, 3.4 cm, 18 mm
   (d) 300 m, 0.5 km, 250 m, 0.22 km, 500 m, 0.45 km
   (e) 250 mm, 20 cm, 0.26 m, 450 mm, 23 cm, 0.5 m

3 Put these weights in order, smallest first.
   (a) 300 g, 0.2 kg, 250 g, 0.4 kg, 500 g
   (b) 4000 g, 4.5 kg, 3.5 kg, 3750 g, 4575 g
   (c) 675 g, 0.7 kg, 0.6 kg, 650 g, 700 g, 0.75 kg

4 Put these volumes in order, smallest first.
   (a) 200 m*l*, 300 m*l*, 0.25 *l*, 0.275 *l*, 260 m*l*
   (b) 1 *l*, 900 m*l*, 1100 m*l*, 1.2 *l*, 1250 m*l*, 0.75 *l*

Exercise 13.3   Links: (*13E, F*) 13E, F

Remember:

Metric	Imperial
8 km	5 miles
1 kg	2.2 pounds
25 g	1 ounce
1 *l*	$1\frac{3}{4}$ pints
4.5 *l*	1 gallon
1 m	39 inches
30 cm	1 foot
2.5 cm	1 inch

The Williams family go on holiday to Devon. The Williams family consist of Mr and Mrs Williams and their three children Lee, Siân and Anna.

1 Mr Williams works out the distance from their home in Swindon to Devon. He makes it 200 miles. What is this distance in kilometres?

2 Mrs Williams buys two 2-litre bottles of cola for the trip. How many pints is this?

3 Anna eats 4 ounces of sweets on the trip. How many grams is this? Write this amount in kilos.

4 Siân buys a 500 gram bar of chocolate. How many pounds is that?

5 Mr Williams fills the car up with 20 litres of petrol. How many gallons is that?

6 Mrs Williams puts 500 ml of oil in the engine. How many pints is that?

7 The family stop at a service station 200 kilometres from home. How many miles is that?

8 Mr Williams puts 2 pints of water in the radiator. How many litres is that?

9 The luggage weighs approximately 150 kg. How many pounds is that?

10 Mrs Williams' luggage weighs 100 pounds. How many kilograms is that?

11 Lee cycles the last 30 miles. How many kilometres is that?

12 Each case is 1 m long by 60 cm wide by 30 cm deep. Change these measurements to inches.

Exercise 13.4   Links: (*13G, H*) 13G, H

1 Change these times into minutes.
(a) 3 hours   (b) 4 hours   (c) $2\frac{1}{2}$ hours   (d) $4\frac{1}{4}$ hours
(e) $\frac{1}{2}$ hour   (f) $\frac{3}{4}$ hour   (g) 3 hours 35 minutes

2 Change these times into hours.
(a) 120 minutes   (b) 4 days   (c) 30 minutes
(d) 300 minutes   (e) $2\frac{1}{2}$ days   (f) 200 minutes
(g) 15 minutes   (h) 2 days   (i) 150 minutes

3 How many seconds are there in
(a) 5 minutes   (b) $\frac{1}{2}$ minute   (c) $\frac{1}{4}$ minute
(d) 8 minutes   (e) 1 hour   (f) $2\frac{1}{2}$ hours

4 Add 20 minutes to each of these times.
(a) 09:00   (b) 10:20   (c) 11:45   (d) 12:50

5 Add 45 minutes to each of these times.
(a) 09:00   (b) 10:20   (c) 11:45   (d) 12:50

6 Add 3 hours 20 minutes to each of these times.
   (a) 10:00   **(b)** 12:30   **(c)** 15:40   **(d)** 16:55

7 Subtract 10 minutes from each of these times.
   (a) 10:40   **(b)** 12:00   **(c)** 10:10   **(d)** 09:05

8 Subtract 45 minutes from each of these times.
   (a) 13:55   **(b)** 16:45   **(c)** 10:30   **(d)** 08:15

9 Subtract 2 hours 30 minutes to each of these times.
   (a) 15:50   **(b)** 16:30   **(c)** 14:10   **(d)** 07:05

10 Alice arrives home at 16:30. She watches television for $2\frac{1}{4}$ hours then spends 1 hour 40 minutes on her homework.
   (a) At what time does Alice start her homework?
   (b) At what time does Alice finish her homework?

Exercise 13.5   Links: (*13I*) 13I

Use this part of a calendar to answer these questions.

Day	June	July
Sunday	5 12 19 26	3 10 17 24 31
Monday	6 13 20 27	4 11 18 25
Tuesday	7 14 21 28	5 12 19 26
Wednesday	1 8 15 22 29	6 13 20 27
Thursday	2 9 16 23 30	7 14 21 28
Friday	3 10 17 24	1 8 15 22 29
Saturday	4 11 18 25	2 9 16 23 30

1 What day of the week is the 16th June?
2 Which day and date is 3 days after the 3rd of June?
3 Which day and date is 10 days before the 7th of July?
4 What is the day and date 14 days after the 20th June?
5 What is the day and date a week before the 2nd July?
6 What is the date 2 weeks after the 17th June?
7 What is the date 10 days after the 31st July?

8 Count on 8 days from the following dates.
   (a) 4th Jan   **(b)** 3rd Feb   **(c)** 7th May
   (d) 25th Aug   **(e)** 25th Sept   **(f)** 25th Dec

9 Count back 15 days from the following dates.
   (a) 21st Oct   **(b)** 16th July   **(c)** 29th Feb
   (d) 8th July   **(e)** 7th May   **(f)** 6th Jan

> **Remember:**
> 30 days hath September
> April June and November.
> All the rest have 31
> except for February alone
> which has just 28 days clear
> and 29 in each leap year.

Exercise 13.6   Links: (*13J*) 13J

Bus Timetable			
Bus station	08:00	09:15	10:30
Stadium	08:15	09:30	10:45
High St.	08:35	09:50	11:05
Hospital	08:45	10:00	11:15
Museum	08:50	10:05	11:20
Bus station	09:10	10:25	11:40

Train Timetable			
Swindon	08:00	09:30	12:45
Kemble	08:15	09:45	13:00
Stroud	08:28	09:58	13:13
Stonehouse	08:40	10:10	13:25
Gloucester	08:55	10:25	13:40
Cheltenham	09:05	10:35	13:50

Use the two timetables above to answer these questions.

1 At what time should the 08:00 bus from the bus station arrive at the High St?

2 At what time should the 08:00 train from Swindon arrive in Gloucester?

3 At what time does the 10:05 bus from the museum leave the stadium?

4 At what time does the 9:58 train from Stroud leave Kemble?

5 Buses from the bus station leave every 1 hour 15 minutes. Continue the bus timetable for the next 3 buses. You may assume that each bus takes the same amount of time between stops as the 08:00 bus.

6 The next 2 trains from Swindon leave at 14:15 and 16:05. Continue the train timetable for these next 2 trains. You may assume that each train takes the same amount of time between stops as the 08:00 train.

7 Steven arrives at the train station in Kemble at 09:40. What time is the next train he could catch to Stroud?

8 Florence arrives at the bus stop in the High St at 11:00. What time is the next bus she could catch to the museum?

9 Work out how long it should take to travel between
   (a) the High St and the museum
   (b) the hospital and the bus station
   (c) the stadium and the museum
   (d) Kemble and Stonehouse
   (e) Stroud and Gloucester.

10 The bus timetable is for a bus route in Gloucester. It takes seven minutes to walk from the bus station to the train platform. Use both timetables to work out which bus and train:
(a) Henry catches at the museum to be in Cheltenham by 10:45
(b) Phil catches at the hospital to be in Cheltenham by 14:00
(c) Sue catches at the stadium to be in Cheltenham for 17:30.

Exercise 13.7   Links: (*13K*) 13K

1 Jake and his family are travelling along the Auto route to the South of France. They see this sign on the side of the road.
(a) How many miles are they from Paris?
(b) How many miles are they from Lyon?
(c) How many miles are they from Marseille?

All the three cities are joined in order by the same road.
(d) How many miles are there from Lyon to Marseille?
(e) How many miles are there from Paris to Marseille?

Auto route to the South	
Paris	80 Km
Lyon	480 Km
Marseille	720 Km

2 Sarah cuts 3 pieces of wood from one 3 metre length. The 3 pieces are of length 1.2 m, 50 cm and 75 cm. How much wood is left? (You may ignore the width of the saw cuts).

3 Susan is making curtains. She buys 8 metres of plain material. She cuts 4 lengths each 1.60 metres long. How much material is left?

4 Put these lengths of wood in order, smallest first

2.8 m,   3 m,   270 cm,   2650 mm,   2590 mm

5 Here is a bus timetable for journeys from Cardiff to London.
(a) At what time should the bus reach
     (i) Bath   (ii) Reading?
(b) How long does it take to get from
     (i) Cardiff to London,
     (ii) Cardiff to Swindon,
     (iii) Bath to Reading,
     (iv) Swindon to Heathrow Airport,
     (v) Bath to London?
(c) The next two buses leave at 10:00 and 12:15. Write out the timetable for these two buses. You may assume that the buses take the same time between the same places.

Place	Time
Cardiff	08:00
Newport	08:25
Bath	09:45
Swindon	10:30
Reading	11:30
Heathrow Airport	12:10
London	12:45

6 Bill's truck is capable of carrying 10 tonnes. How many bags will he be able to carry if each bag weighs
(i) 25 kg   (ii) 50 kg   (iii) 40 kg?

14 Percentages

1 The large square is divided into 100 small squares:

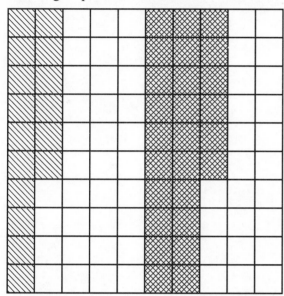

(a) What percentage of the square is shaded?

(i)   (ii)

(b) What percentage is unshaded?

(c) What fraction is shaded?

(d) Write as decimals the amounts of the large square that are:

(i) shaded

(ii) shaded

(iii) unshaded.

2 Change these percentages to fractions in their simplest form.
   (a) 30%   (b) 64%   (c) 57%   (d) 82%   (e) 75%
   (f) $17\frac{1}{2}\%$ (g) $33\frac{1}{3}\%$ (h) 4%   (i) 15%   (j) $\frac{1}{4}\%$
   (k) 28.5%   (l) 3.25%   (m) $5\frac{1}{2}\%$ (n) 82.3%   (o) 6.8%

3 Convert these percentages to decimals.
   (a) 40%   (b) 87%   (c) 15%   (d) 4%   (e) 27.8%
   (f) 7.8%   (g) $22\frac{1}{2}\%$ (h) 0.8%   (i) $\frac{1}{4}\%$ (j) $17\frac{1}{2}\%$
   (k) 58%   (l) 27.5%   (m) 62.4%   (n) 180%   (o) 258%

4 Copy and complete this table of equivalent percentages, decimals and fractions.

Percentage	Decimal	Fraction
60%	0.6	$\frac{3}{5}$
50%		
	0.73	
85%		
$27\frac{1}{2}\%$		
	0.25	
		$\frac{2}{5}$
		$\frac{17}{25}$
	0.08	
$2\frac{1}{3}\%$		
		$\frac{1}{3}$

5 Change these decimals to percentages:
 (a) 0.5   (b) 0.75   (c) 0.25
 (d) 0.68   (e) 0.39   (f) 0.41
 (g) 0.3   (h) 0.72   (i) 0.15
 (j) 0.99   (k) 0.86   (l) 0.31
 (m) 0.065   (n) 0.006   (o) 0.001

6 Write each of these fractions as a percentage:
 (a) $\frac{3}{4}$ (b) $\frac{1}{4}$ (c) $\frac{1}{2}$ (d) $\frac{2}{5}$ (e) $\frac{3}{5}$
 (f) $\frac{7}{10}$ (g) $\frac{9}{20}$ (h) $\frac{17}{25}$ (i) $\frac{1}{8}$ (j) $\frac{5}{8}$
 (k) $\frac{1}{16}$ (l) $\frac{17}{1000}$ (m) $\frac{3}{50}$ (n) $\frac{2}{25}$ (o) $\frac{5}{500}$

7 Re-arrange these numbers in order of size starting with the smallest:
 (a) 53%, 0.52, $\frac{9}{15}$ (b) 83%, $\frac{4}{5}$, 0.82
 (c) 0.006, $\frac{6}{100}$, 6.1%   (d) $\frac{4}{5}$, 80%, 0.8

Exercise 14.2   Links: *(14D–G)* 14E–H

1 Work out:
 (a) 40% of 275   (b) 15% of £800   (c) 12% of 800 m
 (d) 8% of £450   (e) 10% of £368   (f) 3% of 6000
 (g) 17.5% of £2800   (h) 7.8% of 3500 tonnes

2 The total mark in an examination was 80. The pass mark was set at 60% of the total mark. Work out the pass mark.

3 John invests £450 in a building society. At the end of one year the building society pays John 6.5% interest on the investment. How much interest will John be paid at the end of one year?

4 The value of a new motor cycle is £3500. When it is 3 years old the motor cycle will have lost 52% of its value when new.
  (a) How much of its value when new will the motor cycle have lost when it is 3 years old?
  (b) Work out the value of the motor cycle when it is 3 years old.

5 **(a)** What percentage of £200 is £24?
  (b) What percentage of 1500 m is 300 m?
  (c) Write £600 as a percentage of £750.
  (d) Write 40 cm as a percentage of 85 cm.

6 The normal price of a coat is £60.
The sale price of this coat is £36.
Write the sale price as a percentage of the normal price.

7 Glassport's employ 1650 people. 395 of these people are left-handed. What percentage of the people employed by Glassport's are left-handed? Give your answer to the nearest whole number.

8 Frances sees three different advertisements for jeans.

Work out the cost of the jeans in each advertisement.   [E]

9 An estate agency charges $1\frac{1}{2}\%$ commission on the selling price of a house. Work out the commission on a house with a selling price of
(**a**) £60 000   (**b**) £125 000   (**c**) £270 000

10 A van is advertised for sale. The advertisement reads:

> *For Sale*   £3200 plus VAT

VAT is charged at 17.5%.
(**a**) Work out the VAT that will be charged on the van.
(**b**) Work out the total selling price of the van when VAT is added.

11 Asif paid a deposit of £384 on a new motor cycle with a cost price of £3200. What percentage of the cost price was the deposit?

12 In a sale, the price of a coat was reduced from £110 to £82.50. What was the percentage reduction in the price of the coat?

13 James invests £400 in a building society for one year. The building society pays an interest rate of 6.5% per year.
(**a**) How much interest, in pounds, does James receive?

His friend, Megan, invests £600 in another building society for a year. At the end of the year she receives £32.40 interest.
(**b**) Work out the percentage interest rate for this second building society.

14

Equate Computer System	
Normal Price	£1600
Sale Price	£1360

Work out the percentage reduction on the Equate Computer System in the sale.

15 Manuel received a pay rise of 4.5%. Before the pay rise his weekly pay was £240. Work out his weekly pay after the pay rise.

16 Write:
(**a**) 16 as a percentage of 40   (**b**) 12 as a percentage of 30
(**c**) £300 as a percentage of £500   (**d**) £250 as a percentage of £400.

17 One mile = 1600 metres (approximately).
   Write one kilometre as a percentage of a mile.

18 In a sale, the price of a pair of shoes was reduced from £60 to
   £51. Work out the reduction as a percentage of the original
   price.

19 Jim and Leslie bought a flat for £65,000. They sold it a year later
   for £76,700. Work out the profit they made as a percentage of
   the amount they paid for the flat.

20 In 1961 the average price of a litre of petrol was 2.5p.
   In 2001 the average price of a litre of petrol was 80p.
   Work out the percentage increase in the price of a litre of petrol
   over the 40 years.

15 Algebra 3

1 Find the value of the letter in these equations:

(a) $a + 3 = 7$ (b) $b + 2 = 6$ (c) $c + 4 = 5$

(d) $w - 2 = 6$ (e) $x + 5 = 9$ (f) $y - 3 = 5$

(g) $d + 6 = 11$ (h) $m - 6 = 7$ (i) $10 + x = 19$

(j) $15 = y + 3$ (k) $x - 6 = 0$ (l) $8 - x = 5$

(m) $p + 9 = 16$ (n) $r - 3 = 7$ (o) $a + 4 = 9$

(p) $b - 3 = 4$ (q) $d + 1 = 14$ (r) $10 - f = 6$

2 Find the value of the letter in these equations.
You may find the balancing method helps you:

(a) $x + 5 = 9$ (b) $a + 2 = 7$ (c) $c + 1 = 8$

(d) $t - 4 = 3$ (e) $m - 5 = 8$ (f) $n - 8 = 9$

(g) $y + 5 = 5$ (h) $s + 5 = 7$ (i) $x - 9 = 0$

(j) $d + 4 = 15$ (k) $f - 3 = 11$ (l) $4 - y = 2$

(m) $10 = x + 2$ (n) $15 - x = 7$ (o) $x - 18 = 25$

(p) $8 = y + 8$ (q) $t - 7 = 10$ (r) $8 = r - 3$

(s) $15 = 8 + x$ (t) $22 = 12 + x$ (u) $11 = x - 11$

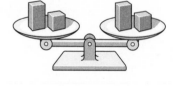

Find the value of the letter in each of these equations:

1 $2x = 4$ **2** $3y = 9$ **3** $2a = 6$

4 $4y = 12$ **5** $5c = 20$ **6** $3y = 18$

7 $9t = 18$ **8** $4w = 24$ **9** $7x = 35$

10 $3x = 10$ **11** $\dfrac{x}{2} = 4$ **12** $\dfrac{y}{3} = 12$

13 $\dfrac{a}{7} = 3$ **14** $\dfrac{b}{4} = 5$ **15** $\dfrac{c}{7} = 1$

16 $\dfrac{t}{4} = 6$ **17** $\dfrac{a}{11} = 5$ **18** $\dfrac{y}{2} = 13$

19 $\dfrac{f}{4} = 4$ **20** $\dfrac{m}{9} = 3$ **21** $x + 9 = 18$

22 $b - 4 = 13$ **23** $15 - x = 1$ **24** $7g = 42$

25 $\dfrac{a}{5} = 5$ **26** $x + 11 = 25$ **27** $y - 7 = 18$

28 $r - 7 = 0$ **29** $3 - x = 11$ **30** $15 = y + 12$

31 $9x = 72$ **32** $15 = \dfrac{a}{3}$ **33** $45 = 5x$

34 $11 + t = 19$ **35** $11 - s = 16$ **36** $6p = 24$

Exercise 15.3   Links: (*15H, I, J*) 15G, H, I, J

Find the value of the letter in each of these equations:

1 $2x + 1 = 5$ **2** $2x - 3 = 5$ **3** $2x + 5 = 7$ **4** $3y + 4 = 13$
5 $4a - 5 = 11$ **6** $5t + 1 = 26$ **7** $6m - 3 = 18$ **8** $9x + 6 = 33$
9 $4a + 1 = 11$ **10** $2b + 5 = 1$ **11** $3d + 4 = 1$ **12** $4x + 9 = 5$
13 $5t + 11 = 7$ **14** $2x + 7 = 3$ **15** $3x - 2 = -14$ **16** $3y + 7 = 2$
17 $4x - 10 = -10$ **18** $2a + 15 = 3$ **19** $6s + 9 = 30$ **20** $4 + 2x = 8$
21 $10 + 2y = 6$ **22** $12 = 2x + 5$ **23** $6k - 9 = 15$ **24** $3x - 7 = -11$
25 $12x + 5 = 5$ **26** $9y + 9 = 0$ **27** $15a + 1 = 46$ **28** $4t - 13 = 19$
29 $3g + 11 = 23$ **30** $5x + 7 = 31$ **31** $\frac{x}{2} + 3 = 5$ **32** $\frac{y}{3} - 7 = 1$
33 $\frac{r}{3} - 1 = -4$ **34** $\frac{x}{7} - 1 = 2$ **35** $\frac{a}{3} + 2 = 5$ **36** $-\frac{c}{4} + 1 = 3$

Exercise 15.4   Links: (*15K, L*) 15K, L

Solve these equations:

1 $2(x + 3) = 10$ **2** $2(y - 3) = 6$ **3** $3(a + 4) = 21$ **4** $4(d - 4) = 12$
5 $3(t + 5) = 24$ **6** $2(w - 5) = 8$ **7** $7(x + 1) = 21$ **8** $4(x + 5) = 12$
9 $5(d + 1) = 0$ **10** $3(2d + 1) = 9$ **11** $2(3x - 2) = 14$ **12** $2(5y - 3) = 14$
13 $2(5p - 2) = 46$ **14** $4(2x - 1) = 3$ **15** $3(4y - 3) = 7$

Exercise 15.5   Links: (*15M, N*) 15M, N

Solve these equations:

1 $2x + 3 = x + 7$ **2** $5x - 3 = 4x + 1$ **3** $5y + 7 = 3y - 5$ **4** $4y - 3 = 3y + 1$
5 $12t - 8 = 7t + 7$ **6** $3x - 8 = x + 4$ **7** $5a + 3 = 3a - 4$ **8** $2b + 7 = 8b - 3$
9 $11c - 9 = 4c + 3$ **10** $5x - 2 = 3x - 7$ **11** $8y - 9 = 5y + 6$ **12** $4s - 3 = 7s - 15$
13 $3a - 2 = a + 4$ **14** $3x + 5 = 5x + 9$ **15** $5x - 9 = 2x + 1$ **16** $4y - 11 = 7y + 4$
17 $3s + 8 = 7s - 9$ **18** $8g + 1 = 15g - 6$ **19** $9y + 15 = 7y - 3$ **20** $4x + 5 = 3x - 4$

Exercise 15.6   Links: (*15O, P*) 15O, P

Solve these equations:

1 $2(x + 1) = 3(2x - 4)$ **2** $4(3t - 2) = 5(2t - 3)$
3 $3(2s + 3) = 2(5s - 8)$ **4** $5(2a - 7) = 4(2a + 3)$
5 $4(k - 3) = 5(2k + 3)$ **6** $3(4x - 5) = 3(3x + 2)$
7 $3x + 11 = 5x + 3$ **8** $6x - 5 = 3x + 4$
9 $9x - 7 = 11x + 3$ **10** $4(5k - 2) = 32$
11 $4(a + 5) = 5(a - 2)$ **12** $20 = 6(2x + 3)$
13 $3(p + 4) = 2(p + 1)$ **14** $7x + 9 = 5x - 16$
15 $15(2x - 1) = 10(2x + 5)$ **16** $2(2x + 3) + 3(x - 1) = 4(x - 4)$

16 Pie charts

1 Here are the number of animals Bill has on his farm.

Animal	Frequency	Number of degrees
Cows	100	200°
Sheep	50	
Hens	20	
Pigs	10	
Total	180	360°

  (a) Copy and complete the table to show the number of degrees
    for each animal.
  (b) Draw a pie chart to show the data.

2 Bob only has 36 animals on his farm.

Animal	Frequency	Number of degrees
Cows	4	
Sheep	20	
Hens	10	
Pigs	2	
Total	36	360°

  (a) Copy and complete the table to show the number of degrees
    for each animal.
  (b) Draw a pie chart to show the data.

3 Cherry collects data about the favourite colours of the 30
  students in her class. Here are the results.

Colour	Frequency	Number of degrees
Red	7	
Yellow	6	
Green	6	
Blue	5	
Orange	6	
Total	30	360°

  (a) Copy and complete the table to show the number of degrees
    for each colour.
  (b) Draw a pie chart to show the data.

4 Thelma has collected data about the ways the students travelled
to school one day.

Method of travel	Frequency	Number of degrees
Walked	6	
Caught a bus	4	
Came by car	7	
Cycled	3	
Total	20	360°

(a) Copy and complete the table to show the number of degrees
for each method of transport.

(b) Draw a pie chart to show the data.

5 Sandy asked 90 people who stayed for a school dinner what they
ate. Here are the results.

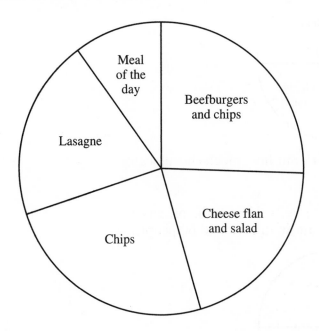

Complete the following table to work out how many students
ate each type of meal. You will need a protractor.

Meal	Number of degrees	Number of students
Meal of the day		
Beefburgers and chips		
Cheese flan and salad		
Chips		
Lasagne		
Total	360°	90

6 Natasha has a take home pay of £180 each week. She allocates her money in the following way.

Allocation	Amount	Number of degrees
Rent	£60	
Travel	£10	
Clothes	£20	
Food	£40	
Savings	£15	
Spare	£35	
Total	£180	360°

Draw a pie chart to show how Natasha allocates her money.

7 Debbie allocates her money in the following way.

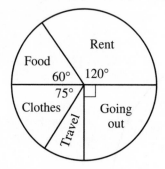

Debbie earns £180 per week. Work out how much she allocates to each item.

8 Tamsin collects data on favourite sweets. Here is the pie chart she produces to show the data. Tamsin interviewed 60 people.

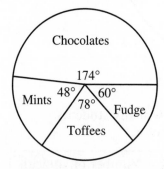

(a) How many liked fudge?
(b) Which was the most popular sweet?
(c) Which was the least favourite sweet?
(d) How many people liked toffees?
(e) How many more people liked chocolates than toffees?

17 Ratio and proportion

1 Here are some tile patterns. For each one write the ratio of white tiles to black tiles:

(a) 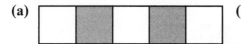   **(b)** 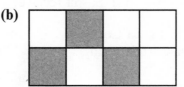   **(c)**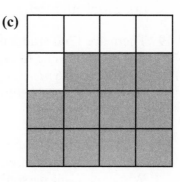

2 Draw a copy of the diagram opposite:

Shade your diagram so that 3 parts of it are shaded and the rest is unshaded.

3 A recipe for 4 cakes needs 50 g of butter and 120 g of flour.
Work out:
(a) how much butter is needed for 8 cakes
(b) how much flour is needed for 8 cakes
(c) how much flour is needed for 12 cakes
(d) how much butter is needed for 6 cakes
(e) how much butter is needed for 18 cakes
(f) how much flour is needed for 10 cakes
(g) how much butter is needed for 10 cakes.

4 A builder uses 3 buckets of sand and 1 bucket of cement to make concrete.
(a) How many buckets of sand will he need when he uses 12 buckets of cement?
(b) How many buckets of cement will he need when he uses 15 buckets of sand?

5 A drink is made by mixing orange juice and water in the ratio 1 : 4.
How much orange juice would need to be used to make a drink which contained 60 cc of water?

6 Write these ratios in their lowest terms:
    (a) $3:6$ **(b)** $2:10$ **(c)** $4:32$ **(d)** $4:10$
    (e) $12:4$ **(f)** $18:9$ **(g)** $30:12$ **(h)** $35:15$

7 Write as ratios in their lowest terms:
    (a) $20\,p:100\,p$ **(b)** $30\,p:£1$ **(c)** $£5:50\,p$
    (d) 2 metres : 10 cm   **(e)** 5 cm : 1 metre   **(f)** 12 min : 1 hour

8 Elizabeth and Sasha share £40 in the ratio $3:5$. How much should
    (a) Elizabeth receive   **(b)** Sasha receive?

9 There are 1300 students at Lucea College.
The ratio of male to female students at
the college is $6:7$.
How many female students are there
at Lucea College?

10 Mr Mohammed won £3600 on the Lottery. He shared the
money between himself, his wife and his son in the ratio $1:3:2$.
Work out each person's share of the money.

11 Mortar is made by mixing weights of sand and cement in the
ratio $7:2$.
    (a) How much sand is needed to make 1800 kg of mortar?
    (b) How much cement is needed to make 4500 kg of mortar?

Exercise 17.2   **Links:** (*17C, D*) 17C, D

1 Calculate the missing numbers in these ratios:
    (a) $4:5=12:?$ **(b)** $5:8=30:?$ **(c)** $2:7=6:?$
    (d) $?:3=7:6$ **(e)** $?:7=16:28$ **(f)** $?:1=16:10$

2 The diagram represents a lamp-post and the shadow it casts at noon.

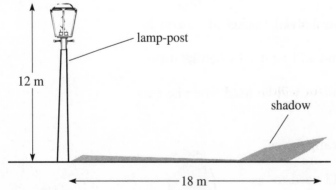

On the same day at noon:
    (a) how long is the shadow cast by a tree which is 20 metres tall
    (b) how long is the shadow cast by a man who is 2 metres tall
    (c) how tall is a building which casts a shadow of length 60 metres?

3 The ratio of the length of a field to its width is 8 : 3. The length of the field is 200 metres. Work out the width of the field.

4 Joan works for 4 hours and is paid £18. Sam earns the same amount for each hour's work as Joan. How much should Sam be paid for 10 hours work?

5 Six men can build a boat in 24 days. Working at the same rate, how long should it take:
(a) 12 men to build the same sort of boat
(b) 9 men to build the same sort of boat?

6 Ten litres of central heating oil cost £1.62. Work out the cost of:
(a) 1 litre of this oil
(b) 100 litres of this oil
(c) 750 litres of this oil

7 Each first class stamp costs 26 p. Work out the cost of:
(a) 10 first class stamps   **(b)** 100 first class stamps
(c) 30 first class stamps   **(d)** 420 first class stamps

8 A packet of 20 Christmas cards costs £3.20.
(a) Work out the cost of each Christmas card.
(b) Work out the cost of 12 of these packets of Christmas cards.

9 Asif works as a sales representative. He receives 35 p for each mile he travels on business as travel expenses. Work out how much he will receive as travel expenses if he travels:
(a) 10 miles   **(b)** 30 miles   **(c)** 250 miles

Last year Asif travelled 8400 miles on business.
(d) How much did he receive as travel expenses?

10 A group of friends hire a boat for a week. The cost of hiring the boat for a week is £560. The group of friends share this cost equally between them.
How much will it cost each person if the boat is:
(a) hired for a week and there are
    (i) 4 people in the group
    (ii) 10 people in the group

(b) hired for 2 weeks and there are
    (i) 7 people in the group
    (ii) 8 people in the group

Exercise 17.3   Links: (*17E*) 17E

1 A map has a scale of 1 : 20 000. What is the distance on the
   ground if the distance on the map is:
   (a) 2 cm   **(b)** 3.5 cm   **(c)** 14.2 cm   **(d)** 25.8 cm

2 An architect draws a plan of a house. The plan is drawn to a
   scale of 1 : 25. Work out:
   (a) the height of the house on the plan given that the true height
      of the house is 12.5 metres
   (b) the length of the house given that the length of the house on
      the plan is 64 cm
   (c) the width of the house on the plan given that the true width
      of the house is 14.6 metres.

3 The real-life distance between Lucea and Presswell is
   24 kilometres. The distance between Lucea and Presswell on a
   map is 6 cm.
   (a) On the map, how many kilometres are represented by 1 cm?
   (b) Work out the real-life distance between two towns which are
      8 cm apart on the map.
   (c) Work out the scale of the map as a ratio in its lowest terms.

4 A model boat has a scale of 1 cm represents 1.2 metres.
   (a) Write this scale as a ratio.
   (b) What is the length of the model if the length of the real boat
      is 24 metres.
   (c) Work out the width of the real boat if the width of the
      model is 14 cm.

5 A model castle has a scale of 1 : 40.
   (a) Work out the height of the dining room in the castle given
      that the height of the dining room in the model is 15.4 cm.
   (b) Work out the length of the drawbridge on the model castle
      given that the length of the real drawbridge is 8 metres.
   (c) The perimeter of the moat around the castle given that the
      perimeter of the moat on the model is 6.48 metres.

18 Symmetry

1 For each design write down how many lines of symmetry there are.

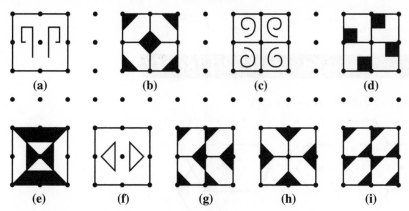

   (a)   (b)   (c)   (d)

   (e)   (f)   (g)   (h)   (i)

2 Copy each shape and draw in all the lines of symmetry.

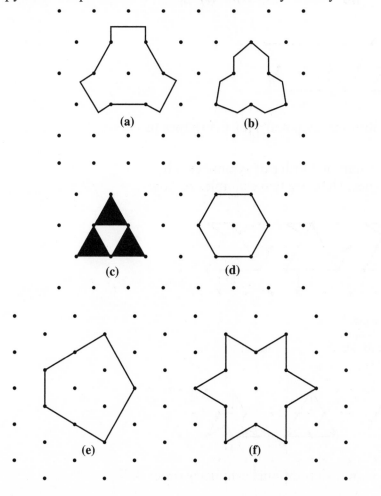

   (a)   (b)

   (c)   (d)

   (e)   (f)

3 Using the dotted lines as lines of symmetry, copy and complete each shape.

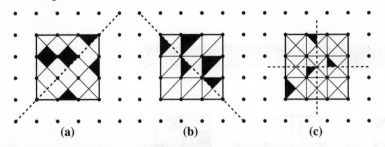

(a)   (b)   (c)

Exercise 18.2   Links: *(18C, D)* 18C, D

1 Which of these shapes have rotational symmetry?

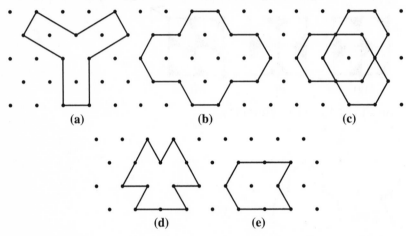

(a)   (b)   (c)

(d)   (e)

2 Write down the order of rotational symmetry for the shapes in question **1**.

3 You can form shapes with rotational order of symmetry 3 by modifying equilateral triangles. Here are two examples of how you can do this.

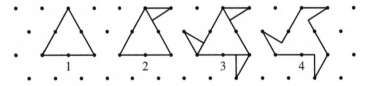

1   2   3   4

1 Start with an equilateral triangle.
2 Add something extra to one side.
3 Add the same to the other two sides.
4 Redraw with only the outline.

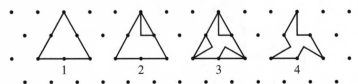

1   2   3   4

Make some shapes of your own with rotational symmetry order 3.

4 Use the same idea with a square to make shapes with order of symmetry 4.

Exercise 18.3   Links: *(18E)* 18E

1 Copy or trace each of these shapes and draw in any planes of symmetry that they have.

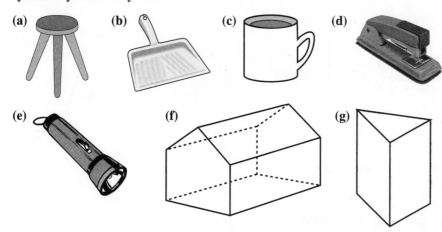

(a)   **(b)**   **(c)**   **(d)**

(e)   **(f)**   **(g)**

2 Copy and complete each shape using the shaded plane as the plane of symmetry.

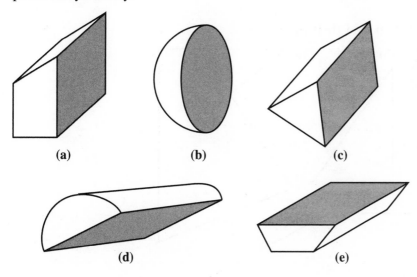

(a)   **(b)**   **(c)**

(d)   **(e)**

19 Measure 3

1 Work out the perimeters of these shapes:

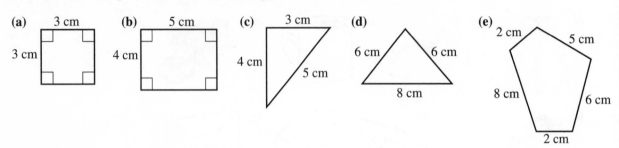

(a) 3 cm, 3 cm
(b) 5 cm, 4 cm
(c) 3 cm, 4 cm, 5 cm
(d) 6 cm, 6 cm, 8 cm
(e) 2 cm, 5 cm, 8 cm, 6 cm, 2 cm

2 Work out the perimeters of these shapes:
   (a) a square with side 6 cm
   (b) a rectangle with sides 6 cm and 5 cm
   (c) an equilateral triangle with sides 10 cm
   (d) a regular pentagon with sides 4 cm
   (e) an isosceles triangle with two sides of 4 cm and one of 6 cm

3 Measure these shapes and then find their perimeters:

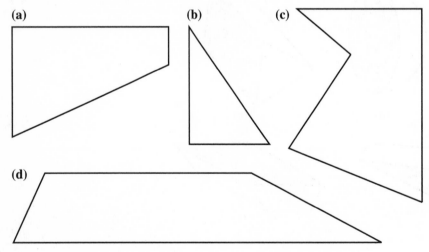

(a)
(b)
(c)
(d)

4 Find the perimeters of these shapes:

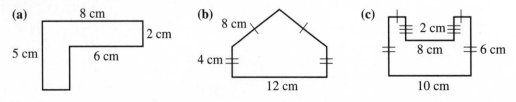

(a) 8 cm, 2 cm, 5 cm, 6 cm
(b) 8 cm, 4 cm, 12 cm
(c) 2 cm, 8 cm, 6 cm, 10 cm

1 Work out the circumferences of these circles with diameters:
  (a) 5 cm   **(b)** 4 cm
  (c) 6 cm   **(d)** 10 cm
  (e) 1.5 m   **(f)** 2.8 m
  (g) 3.6 m   **(h)** 12 m
  (i) 25 cm   **(j)** 1 km
  (k) 2.25 m   **(l)** 5.6 cm

2 Work out the circumferences of these circles with radii:
  (a) 5 m   **(b)** 3 cm
  (c) 2 m   **(d)** 2.5 m
  (e) 3.5 cm   **(f)** 4.5 m
  (g) 2.4 m   **(h)** 5.6 cm
  (i) 20 mm   **(j)** 30 mm
  (k) 2.1 m   **(l)** 1 m

3 A wheelbarrow has a wheel with a diameter of 30 cm. Work out
the circumference of the wheelbarrow wheel.

4 A bicycle has a wheel with a radius of 25 cm. Work out the
circumference of the bicycle wheel.

5 Robin travels 600 metres on a bicycle. The circumference of the
bicycle wheel is 75 cm. How many times does the wheel rotate on
the journey?

6 Sylvie travels 1.2 kilometres on her bicycle. The diameter of the
bicycle wheel is 60 cm.
  (a) Work out the circumference of the bicycle wheel.
  (b) How many times does the wheel rotate on the journey?

7 A tree with a circular trunk has a circumference
of 2.8 m. Work out the diameter of the tree trunk.

2.8 m

8 Copy this table into your book. Work out the diameter and
radius of the circles.

Circumference	Diameter	Radius
(a) 50 cm		
(b) 45 mm		
(c) 10 mm		
(d) 15 cm		
(e) 3.14 m		

9 One of Sven's rollerblade wheels rotates 5000 times when he
travels 600 metres. Work out the diameter of the wheel.

10 The circumference of a circular village pond is 60 metres. Work
out the radius.

Exercise 19.3   Links: *(19H, I, J)* 19H, I, J

1 Find the areas of these shapes by counting squares.

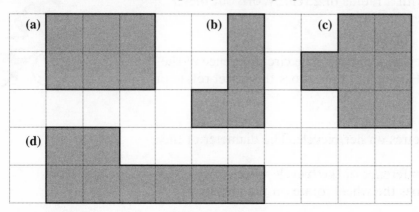

2 Estimate the areas of these shapes by counting squares.

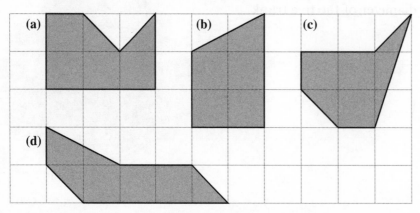

3 Estimate the areas of these curved shapes by counting squares.

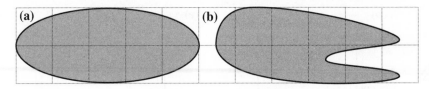

(a) **(b)**

4 Find the volumes of these shapes by counting cubes.

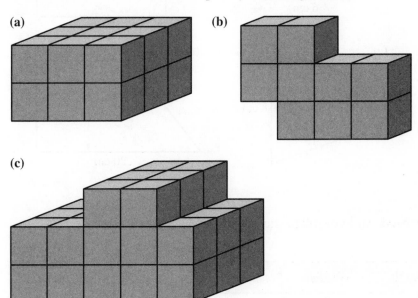

(a) **(b)**

(c)

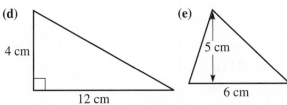

Exercise 19.4   **Links: (*19K, L*) 19K, L**

1 Find the areas of these shapes.

(a) 5 cm / 4 cm

(b) 7 cm / 12 cm

(c) 10 cm / 8 cm

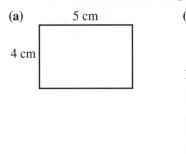

(d) 4 cm / 12 cm

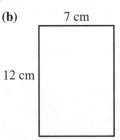

(e) 5 cm / 6 cm

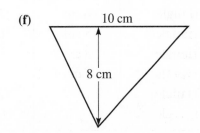

(f) 10 cm / 8 cm

2 Find the areas of these composite shapes.

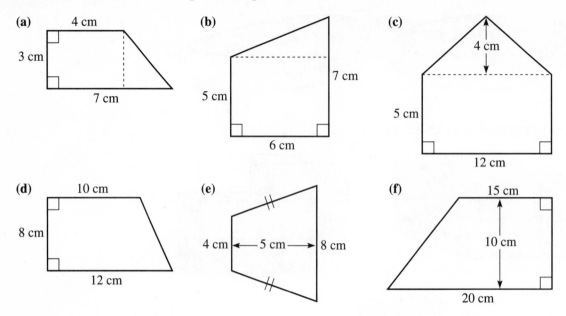

(a) 4 cm, 3 cm, 7 cm

(b) 5 cm, 7 cm, 6 cm

(c) 4 cm, 5 cm, 12 cm

(d) 10 cm, 8 cm, 12 cm

(e) 4 cm ←5 cm→ 8 cm

(f) 15 cm, 10 cm, 20 cm

3 Copy this table into your book and complete the missing numbers.

	Shape	Length	Width	Area
(a)	Rectangle	4 cm	5 cm	
(b)	Rectangle	5 cm	8 cm	
(c)	Rectangle	8 cm		32 cm^2
(d)	Rectangle	7 cm		28 cm^2
(e)	Rectangle		2 cm	16 cm^2
(f)	Rectangle		9 cm	108 cm^2

4 Copy this table into your book and complete the missing numbers.

	Shape	Base	Vertical height	Area
(a)	Triangle	6 cm	5 cm	
(b)	Triangle	8 cm	6 cm	
(c)	Triangle	8 cm		16 cm^2
(d)	Triangle	7 cm		21 cm^2
(e)	Triangle		2 cm	12 cm^2
(f)	Triangle		9 cm	36 cm^2

5 Work out the surface area of a cube with sides of length 5 cm.

Hint: a cube has 6 faces.
Each face is a square.

6 Work out the surface area of these shapes.

(a)

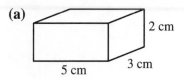

2 cm

3 cm

5 cm

(b)

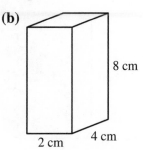

8 cm

2 cm   4 cm

(c)

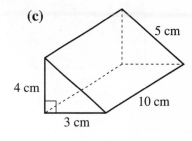

5 cm

4 cm

3 cm

10 cm

Exercise 19.5   Links: (*19M, N, O*) 19M, N, O

1 Find the areas of the circles with radii:
  (a) 5 cm   **(b)** 7 cm   **(c)** 11 cm   **(d)** 4 cm
  (e) 2.5 m   **(f)** 3.2 cm   **(g)** 5.4 m   **(h)** 2.9 m

2 Find the areas of the circles with diameters:
  (a) 8 cm   **(b)** 6 cm   **(c)** 10 cm   **(d)** 18 cm
  (e) 3.2 cm   **(f)** 8.4 cm   **(g)** 6.6 m   **(h)** 12.4 m

3 Work out the area of a:
  (a) circular pond with a radius of 0.9 m
  (b) circular birthday card with a diameter of 120 mm
  (c) crop circle with a radius of 15 m
  (d) pencil with a diameter of 1 cm
  (e) circular mill wheel with a radius of 30 cm.

4 A circle has an area of 10 cm². Work out the radius.

5 A circle has an area of 50 m². Work out the diameter.

6 A circular flower pot has an area of 12 cm². Work out the diameter of the base.

7 A circular table has a base area of 3 m². Work out the diameter of the table.

8 A pencil eraser in the shape of a circle has an area of 5 cm². Work out the radius of the pencil eraser.

Exercise 19.6   Links: *(19P, Q)* 19P, Q

1 Work out the volumes of these cuboids.

(a)
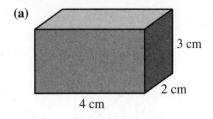
3 cm
2 cm
4 cm

(b)
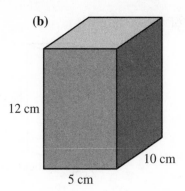
12 cm
5 cm
10 cm

(c)
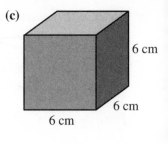
6 cm
6 cm
6 cm

2 Copy the table into your book and fill in the missing numbers.

Shape	Length	Width	Height	Volume
(a) Cuboid	6 cm	4 cm	2 cm	
(b) Cuboid	5 cm	4 cm	3 cm	
(c) Cube	3 cm	3 cm	3 cm	
(d) Cuboid	6 cm		3 cm	36 cm^3
(e) Cuboid	8 cm		2 cm	48 cm^3
(f) Cuboid		4 cm	2 cm	24 cm^3
(g) Cuboid		5 cm	3 cm	75 cm^3
(h) Cuboid	4 cm	2 cm		32 cm^3
(i) Cuboid	10 cm	5 cm		100 cm^3
(j) Cube	4 cm			64 cm^3
(k) Cube				1000 cm^3

3 How many cubes of side 2 cm will fit into a cuboid measuring 10 cm by 8 cm by 6 cm?

4 Bill and Ben decide to put up a greenhouse on a rectangular concrete base. The base for the greenhouse is in the shape of a cuboid which is 2.6 m long, 1.9 m wide and 15 cm deep. Concrete costs £50 per cubic metre. Work out the cost of the concrete base for the greenhouse.

Exercise 19.7   Links: 19R

1 Work out the number of
   (a) cm^2 in 3 m^2 **(b)** cm^2 in 4 m^2
   (c) cm^2 in 3.5 cm^2 **(d)** cm^2 in 10 m^2
   (e) m^2 in 100 000 cm^2 **(f)** m^2 in 30 000 cm^2
   (g) cm^2 in 0.05 m^2 **(h)** m^2 in 1000 cm^2

2 Work out the number of
- **(a)** cm^3 in $3\,m^3$
- **(b)** cm^3 in $5\,m^3$
- **(c)** cm^3 in $2.5\,m^3$
- **(d)** cm^3 in $3.47\,m^3$
- **(e)** m^3 in $6\,000\,000\,cm^3$
- **(f)** m^3 in $50\,000\,000\,cm^3$
- **(g)** cm^3 in $0.05\,m^3$
- **(h)** m^3 in $100\,000\,cm^3$
- **(i)** mm^3 in $1\,m^3$
- **(j)** mm^3 in $5\,cm^3$

3 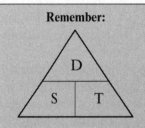 The volume of a garden shed is $6\,m^3$.
How many cm^3 is this?

4 The area of one page in a book is $40\,cm^2$.
There are 100 pages in the book.
Write the total area of paper in m^2.

Exercise 19.8   Links: (*19R*) 19S

> **Remember:**
>
> You can use this triangle to help you remember the formula connecting Distance, Speed and Time

1 Catherine walked for 2 hours at $4\,km$ per hour. How far did she walk?

2 Keith drove for 4 hours at an average speed of 60 miles per hour. How far did he drive?

3 Peter cycled at 12 miles per hour for 3 hours. How far did he cycle?

4 Steve drove 100 kilometres in 2 hours. At what average speed did he drive?

5 Barry walked 15 miles in 5 hours. At what average speed did he walk?

6 Karen travelled 48 kilometres in 3 hours. At what average speed did she travel?

7 Dave drove 120 miles at 60 miles per hour. How long did it take him?

8 Marlene walked 10 miles at 4 miles per hour. How long did it take her?

Exercise 19.9   Links: (*19S*) 19T

1 (a) The diagram shows the plan of a small shop.
The shop is to be carpeted. The cost of the carpet is £10.50 per square metre.
Work out the total cost of the carpet.
(b) Gripper rod is needed to go around the outside of the room. It costs £1.50 per metre. Work out the cost of the gripper rod.

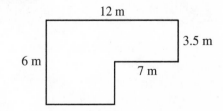

(c) Carpet tiles could also be used to cover the floor. The tiles
   are square with sides of length 50 centimetres. The tiles are
   sold in boxes of 12 for £25 per box.
   (i) Work out the number of boxes needed.
   (ii) Work out the cost of the carpet tiles.

2 Rachael runs a 400 metre race in 2 minutes. Work out her
   average speed in kilometres per hour.

3 Thelma travels at 60 miles per hour for $2\frac{1}{2}$ hours. How far has
   she travelled.

4 Wanda builds a circular fish pond that has a diameter of 4 m.
   (i) Work out the area of the surface of the water.
   (ii) Work out the circumference of the pond.

5 A circular race track has a radius of 250 metres. A horse runs
   once around the track in 10 minutes. Work out the speed of the
   horse. Give your answer in kilometres per hour.

6 Gill uses a jug that is in the shape
   of a cuboid to fill a container that holds
   10 litres. The cuboid has a square base
   with side of length 5 cm and a height
   of 8 cm. How many jugs full will be
   needed to fill the container completely?

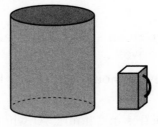

7 Teri cycles 2 kilometres to school. Her bicycle wheel rotates 1050
   times during the journey. Work out the diameter of Teri's
   bicycle wheel. Give your answer to the nearest centimetre.

8 A swimming pool, in the shape of a cuboid, measures 50 m by
   20 m and is 1.8 m deep. Work out how many litres of water
   would be needed to completely fill the swimming pool.

20 Averages

1 Find the mean, median, mode and range of these sets of data:
   (a) 4, 6, 9, 8, 7, 9, 10, 4, 7, 6, 9, 9
   (b) 6, 9, 9, 10, 10, 7, 8, 7, 6
   (c) 5, 5, 5, 5, 5, 5, 5, 5, 5, 5
   (d) 5.5, 5.6, 5.7, 5.3, 5.4, 5.5, 5.2
   (e) 12, 15, 11, 17, 14, 16, 18, 13, 12, 15
   (f) 45, 47, 45, 44, 42, 47, 45, 43, 44
   (g) 65, 67, 65, 64, 62, 67, 65, 63, 64
   (h) 67, 45, 32, 54, 55, 76, 67, 74, 67, 64, 55, 67

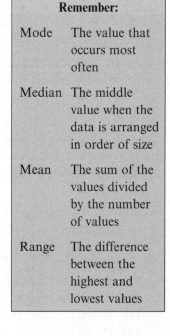

Remember:

Mode	The value that occurs most often
Median	The middle value when the data is arranged in order of size
Mean	The sum of the values divided by the number of values
Range	The difference between the highest and lowest values

2 Myra was preparing a mail-shot and recorded the weights of the packages she was sending. The weights of the first 6 packages were:

   51 g, 55 g, 54 g, 55 g, 56 g, 50 g

   (a) Work out the mean of the data.
   (b) Work out the median value of the data.
   (c) Work out the mode of the data.
   (d) Work out the range of the data.

   Myra then weighed out another 4 packages. Their weights were:

   55 g, 59 g, 60 g, 55 g

   (e) Work out the mean, median and mode of these 4 packages.
   (f) Work out the mean, median, mode and range of all 10 packages.

3 Selena records the marks she obtained in her weekly maths test for one month. Here are the results shown in a bar graph.
   (a) Work out the mean.
   (b) Work out the median.
   (c) Work out the mode.
   (d) Work out the range.

   Selena scored 20 in the next test.
   (e) Work out the mean, median, mode and range of all 5 marks.

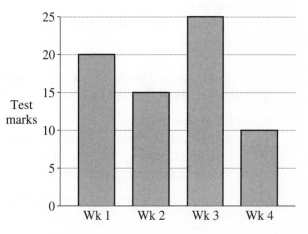

4 Justine measured the heights of the 10 girls in her Maths class. Here are the results:

> 1.80 m, 1.70 m, 1.65 m, 1.72 m, 1.82 m
> 1.65 m, 1.75 m, 1.65 m, 1.74 m, 1.73 m

(a) Work out the mean height.
(b) Work out the median height.
(c) Work out the modal height.
(d) Work out the range of the heights.

Another girl joined the class. When her height was included the mean height increased by one centimetre.
(e) How tall was the new girl?

5 Abbas worked out the mean weight of 10 blocks of cheese. It worked out to be 250 grams. When he added another block of cheese the mean decreased by 1 gram. How heavy was the eleventh block of cheese?

6 Indira checked the weight of nine parcels. Here are the weights in kilograms:

> 1.5, 5.4, 3.0, 3.0, 4.5, 3.0, 7.5, 6.8, 1.4

(a) Work out the mean weight.
(b) Work out the median weight.
(c) Work out the modal weight.
(d) Work out the range of the weights.

Another parcel was added. The mean weight increased by 0.5 kilograms.
(e) What was the weight of the tenth parcel?

7 In this question all the numbers are between 1 and 10. Write down the mean, median, mode and range of a set of six numbers when:
(a) the mean, median and mode are the same number and the range is 0.
(b) the mean, median and mode are the same number and the range is 2.

8 Yousaf calculates the average wage earned by 10 people on a building site. Here are his results:

6 Labourers earning	£200 each
2 Bricklayers earning	£300 each
1 Foreman earning	£350
1 Site Manager earning	£400

Work out the mean, median, mode and range of the data for the 10 people.

Exercise 20.2   **Links:** (*20F*) **20E, F**

1 Calculate the mean, median, mode and range of the data in this table. It gives the weekly wages earned by workers in a small factory.

Worker	Number of workers	Weekly wage	
Machine operator	10	£200	
Chargehand	5	£250	
Foreman	3	£300	
Production manager	1	£500	
Factory owner	1	£1000	
Total	20		

   (a) Copy the table into your book and use the last column to help you work out the mean weekly wage.
   (b) Which average would the factory owner use if he was trying to avoid paying his workers more money? Why?
   (c) Which average would the machine operators use if they wanted to ask for more money? Why?

2 Use this frequency table to work out the mean, median, mode and range of the number of children in a family in Sam's class.

Number of children	Frequency	Frequency × number of children
0	3	0
1	4	4
2	10	20
3	6	
4	4	
5	2	
6	1	
Total	30	

3 Mark bought 5 boxes of apples at £10 a box, 2 boxes of oranges at £15 a box and 1 box of pineapples at £30 a box.
   (a) Work out the mean cost of a box of fruit.
   (b) Which average would Mark use if he wanted to complain about how expensive the fruit was? Why?
   (c) Which average would Mark use if he was boasting to his friends what a good deal he had made? Why?

4 Class 11D measure the time it takes them to run to the top of
the 3-storey classroom block and back down again. Here are the
results. All the times are measured to the nearest second.

Time	Frequency	Frequency × Time
10	1	
11	2	
12	7	
13	10	
14	6	
15	3	
16	1	
Total	30	

Use the data in the table to work out the
(i) mean   **(ii)** median   **(iii)** mode   **(iv)** range

5 Mr Smith recorded for one term the number of absences for his
Year 11 tutor group. Here are the results.

Number of absences	Frequency	Frequency × number of absences
0	2	
2	5	
4	10	
6	4	
8	5	
10	2	
12	1	
30	2	
Total		

(a) Use the data in the table to work out the
(i) mean   **(ii)** median   **(iii)** mode   **(iv)** range
(b) Which average would Mr Smith use if he wanted to
complain about the absence rate in his tutor group? Why?
(c) Which average would Mr Smith use if he wanted to boast
about how little time members of his tutor group had been
absent from school? Why?

Exercise 20.3   Links: 20G

1 The times taken to change an exhaust in Mr Fixit's garage were:

20, 25, 24, 10, 15, 18, 20, 15, 15, 20, 18, 25, 32, 8, 12, 15,
20, 18, 20, 24, 30, 15, 10, 12, 18, 20, 15, 25, 24, 20, 17, 16.

(a) Copy and complete this stem and leaf diagram.

Stem		Frequency
0		
10		
20		
30		

(b) Work out the range of the data.
(c) Work out the mode.
(d) Work out the median.

2 Mr Jones recorded the time it took for his class to complete their maths homework.

5, 10, 12, 25, 15, 16, 10, 8, 9, 19, 10, 6, 8, 9, 12, 15, 20, 16, 9, 12, 6, 10, 12, 15, 12, 20, 25, 20, 15, 22.

(a) Draw a stem and leaf diagram with stem of 10 minutes.
(b) Work out the median.
(c) Work out the mode.

3 The number of pounds lost by members of a slimming club in a 4 week period was

10, 12, 4, 6, 3, 4, 8, 4, 15, 12, 16, 20, 4, 22, 31, 11, 5, 4, 7, 10.

(a) Draw a stem and leaf diagram with a stem of 10 pounds.
(b) Work out the mode.
(c) Work out the range.
(d) Work out the median.

21 Algebra 4

1 To work out her pay Jessica uses the word formula:

   Pay = Rate of pay × hours worked + bonus

   Work out her pay when she works for 25 hours at a rate of pay of £5 an hour and earns a bonus of £8.

2 To work out his pay Pulin uses the word formula:

   Pay = Rate of pay × hours worked + bonus

   Work out his pay when he works for 40 hours at a rate of pay of £4.50 an hour and earns a bonus of £12.

3 Use the formula:

   Cost of golf balls = Cost of one golf ball × number of golf balls

   to work out the cost of 12 golf balls if one golf ball costs £1.60.

4 Write a word formula to help solve each of these problems. Use your formula to work out the answers.
   (a) Trevor works for 35 hours at a rate of pay of £3 an hour. How much should he get paid?
   (b) Arti works for 30 hours at a rate of pay of £4.50 an hour. How much should she get paid?
   (c) Peter buys 15 bars of chocolate at 60 p each. Work out the total cost of the bars of chocolate.
   (d) Clara buys 20 newspapers at 25 p each. Work out the total cost of the newspapers.
   (e) Lucy sold 30 rolls at 45 p each. How much money did she collect?
   (f) Clive sold 60 roses at £3.50 each. How much money did he collect?
   (g) Christine cooked 64 biscuits for a party. Everybody at the party had 2 biscuits. How many people were at the party?
   (h) A window cleaner charges £5.60 to clean 7 windows. He charges the same amount to clean each window. What does he charge to clean 1 window?

1 In this question $a = 2$, $b = 3$, $c = 4$ and $d = 0$. Work out the value of:
   (a) $a + b$ (b) $b + c$ (c) $a + b + c$
   (d) $3a$ (e) abc (f) abd
   (g) $5a - 2b$ (h) $4c + 2a$ (i) $3b - 2a$

(**j**) $a + 3b - 2c$ (**k**) $3c - 2d$ (**l**) $3a + 4b - 4c$
(**m**) $2ab$ (**n**) $ac - ab$ (**o**) $4c - 3a$
(**p**) $3b - a$ (**q**) $7c - 5b + a$ (**r**) $5abc$

2 The formula for the perimeter of a square is $P = 4s$. Find the value of P when:
(**a**) $s = 2$ (**b**) $s = 5$ (**c**) $s = 7$
(**d**) $s = 3.6$ (**e**) $s = 4.8$ (**f**) $s = 16$

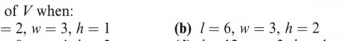

3 The formula for the volume of a cuboid is $V = lwh$. Find the value of V when:
(**a**) $l = 2, w = 3, h = 1$ (**b**) $l = 6, w = 3, h = 2$
(**c**) $l = 8, w = 4, h = 3$ (**d**) $l = 12, w = 3, h = 4$

4 Use the formula $v = u + at$ to work out v when:
(**a**) $u = 2, a = 3, t = 1$ (**b**) $u = 4, a = 4, t = 5$
(**c**) $u = 10, a = 8, t = 9$ (**d**) $u = 8.2, a = 10.2, t = 5$

Exercise 21.3   Links: (*21F–J*) 21F–J

1 Work out:
(**a**) $7 - 4$ (**b**) $5 - 8$ (**c**) $-8 - 5$ (**d**) $4 + (-5)$
(**e**) $-5 - (-3)$ (**f**) $8 - (-4)$ (**g**) $5 - (-6)$ (**h**) $10 + (-12)$

2 In this question $a = -2, b = 3, c = -4, d = 0$. Work out the value of:
(**a**) $a + c$ (**b**) $c + b$ (**c**) $d - c$
(**d**) $c - d$ (**e**) $c - a$ (**f**) $a + b$
(**g**) $b + a$ (**h**) $c + b + a$ (**i**) $c + b - a$
(**j**) $a + b - c + d$ (**k**) $a - c + b$ (**l**) $b - c - a$
(**m**) $d - c + a - b$ (**n**) $b - a - b$ (**o**) $c - a + b$

3 Work out:
(**a**) 3×-2 (**b**) -2×-2 (**c**) -7×5 (**d**) 4×-8
(**e**) -5×-5 (**f**) -8×4 (**g**) 3×-9 (**h**) -6×-4

4 In this question let $a = -4, b = 3, c = -2$ and $d = 0$. Work out the value of:
(**a**) $a + c$ (**b**) $b + c$ (**c**) $a + b + c$
(**d**) $3a$ (**e**) ac (**f**) ab
(**g**) abc (**h**) $5a + 3b$ (**i**) $6b - 2c$
(**j**) $db + bc$ (**k**) $ac + ab - bc$ (**l**) $4ac + 2bc$

5 Use the formula $v = u + at$ to work out v when:
(**a**) $u = 2, a = -3, t = 3$ (**b**) $u = -3, a = 9, t = 5$
(**c**) $u = 5, a = -4.5, t = 4$ (**d**) $u = -10, a = 2, t = 6$

6 Use the formula $m = np + q(r - s)$ to work out m when:
    (a) $n = 2, p = -3, q = 2, r = 5, s = -2$
    (b) $n = -3, p = -4, q = -5, r = 4, s = 3$

Exercise 21.4   Links: 21K

1 Draw this table of values in your book.

x	-3	-2	-1	0	1	2	3
$y = 2x^2 + 3$							

Complete the table by substituting the values of x into the formula to find the values of y.

2 If $a = 3, b = -2$ and $c = 5$ work out the values of:
    (a) $a^2 + c$ **(b)** $c^2 - a$
    (c) $a^2 - b$ **(d)** $a^2 + (a^2 + c)$
    (e) $2(b^2 + c^2)$ **(f)** $3(a + c)^2$
    (g) $(a + c)^2 + (a + b)^2$ **(h)** $3(a + c)^2 - 2(c + b)^2$
    (i) $\sqrt{(c^2 + 3ab^2)}$

Exercise 21.5   Links: (*21K*) 21L

1 Use the information in these diagrams to find the values of the letters:

(a)

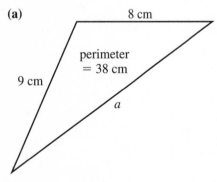

(b)

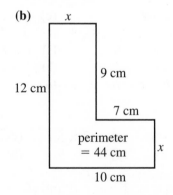

(c)

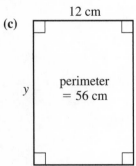

(d)

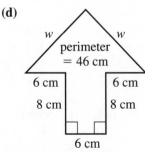

2 Joanna thought of a number. She multiplied it by 3 then added 5. The answer was 17. What number did she first think of?

3 Scott thought of a number. He took away 6 from it, then he multiplied the answer by 5 to get a final answer of 15. What number did Scott think of first?

4 Catherine thought of a number, added 9 to it then multiplied the answer by 4. This gave an answer of 44. What number did Catherine think of?

5 Seamus thought of a number, took 8 away and then divided his answer by 2. The final answer was 16. What number did he think of?

Exercise 21.6   Links: 21M, N

1 Put the correct sign between these pairs of numbers to make a true statement:
   (a) 5, 7   **(b)** 7, 3   **(c)** 5, 5
   (d) 0.2, 0.4   **(e)** 0.1, 0   **(f)** 3.16, 3.15

2 Write down the value of x that are whole numbers and satisfy these inequalities:
   (a) $x > 1$ and $x < 4$ **(b)** $x > 5$ and $x < 10$
   (c) $x < 5$ and $x > 1$ **(d)** $x > 0$ and $x < 7$
   (e) $x > 3$ and $x < 6$ **(f)** $x < 9$ and $x > 5$

3 Draw number lines from 0 to 10. Shade in the inequalities:
   (a) $x > 4$ **(b)** $x < 7$
   (c) $x > 3$ and $x < 6$ **(d)** $x > 4$ and $x < 9$
   (e) $x > 1$ and $x < 4$ **(f)** $5 < x < 8$
   (g) $8 < x < 10$ **(h)** $4 < x < 6$

4 Write down the inequalities represented by the shading of these number lines:

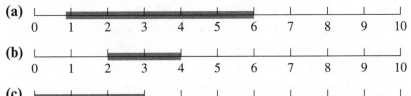

22 Transformations

1 Draw each shape on squared paper and translate it by the
   amount shown.

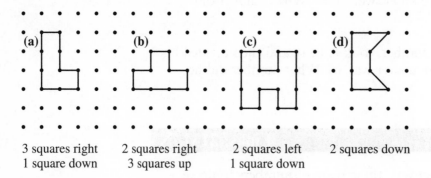

(a)	(b)	(c)	(d)
3 squares right	2 squares right	2 squares left	2 squares down
1 square down	3 squares up	1 square down	

2 Copy the shapes on to squared paper and draw the whole image
   to match the point that has been translated.

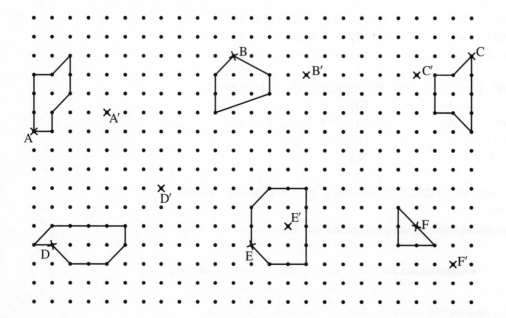

3 Describe the translations for question 2.

Exercise 22.2   Links: (*22B*) 22B

1 Draw separate images for each shape after a rotation of 90°
   anticlockwise for each of the centres marked.

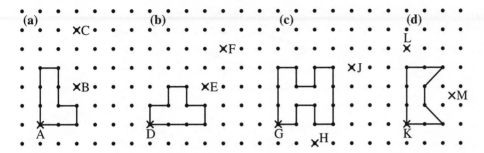

2 Using the same shapes and centres as for question **1**, rotate by a
   half turn.

Exercise 22.3   Links: (*22C*) 22C

For the following questions, reflect each shape in the mirror line A.
On the same diagram, reflect the original shape in the mirror line B.

1

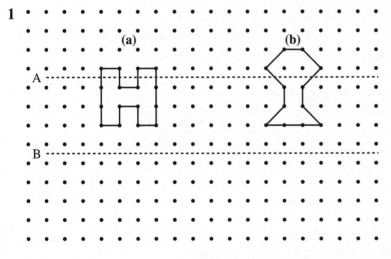

2

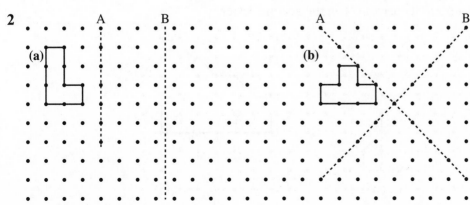

3

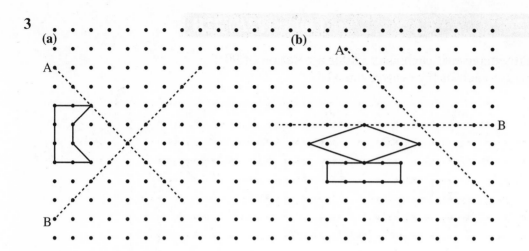

1 In the diagram each shape has two possible centres of
enlargement marked. Use a scale factor of 2 for parts **(a)** and **(c)**
and a scale factor of 3 for part **(b)**.

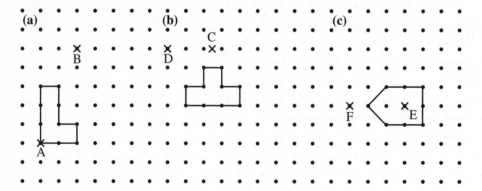

2 Each shape in the diagram has two possible centres of
enlargement marked by two letters. Use a scale factor of 2 for
the first letter and a scale factor of 3 for the second letter.

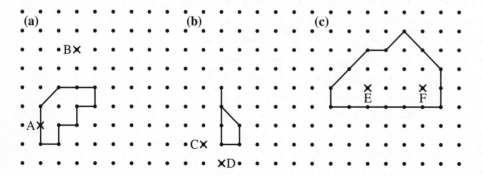

Exercise 22.5   Links: (*22E*) 22E

1   Describe as fully as possible the transformation that takes shape
    1 to shape 2.

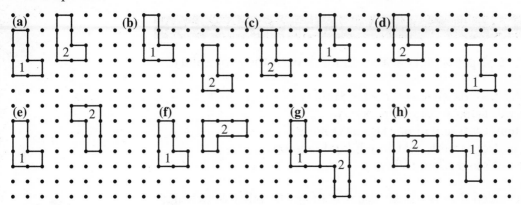

2   Describe as fully as possible the transformation that takes the
    shaded shape to the unshaded shape.

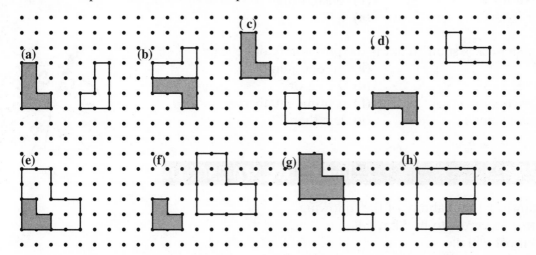

3   For each shape describe as fully as possible the transformation
    which takes shape 1 to shape 2. There may be more than one
    answer.

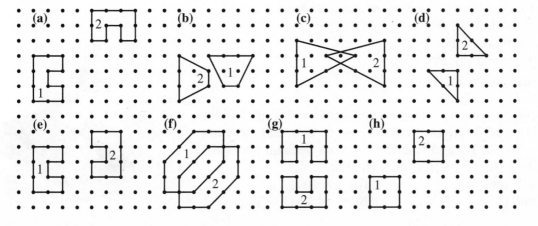

23 Probability

1 Draw a 0 to 1 probability scale and mark on it the probability that:
   (a) Easter Monday follows Easter Sunday
   (b) you will listen to the radio today
   (c) the temperature will not rise next month
   (d) if you drop a table mat it will land face up
   (e) you will see a robin today.

2 A fair six-sided dice is rolled. What is the probability of getting:
   (a) a 3
   (b) a 4 or 5
   (c) a number less than 3
   (d) a 0
   (e) a 1 or 6
   (f) greater than 5
   (g) 1 or 2 or 4
   (h) a 7?

3 The fair spinner is spun. What is the probability of getting:
   (a) a 2
   (b) an odd number
   (c) 3 or 8
   (d) 7 or 4 or 5
   (e) an even number
   (f) a prime number
   (g) greater than 4
   (h) less than 3?

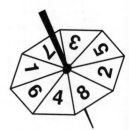

1 Write down the probability of the following:
   (a) A tree will grow to be 5 km tall.
   (b) Your pet will die.
   (c) If you toss a coin it will land on its edge.
   (d) Your pet will live to be 250 years old.
   (e) If you roll a dice it will show an even number.

2 Write two statements for each of the following:
   (a) a probability of 0
   (b) a probability of 1
   (c) a probability of about $\frac{1}{2}$
   (d) a probability of about $\frac{1}{4}$

3 A box of chocolates contains 4 soft centres, 3 nuts, 1 plain chocolate and 2 toffees. Freda takes one chocolate without looking. Write (i) as a fraction (ii) as a decimal and (iii) as a percentage the probability that she takes:
   (a) a soft centre
   (b) a toffee
   (c) a nut
   (d) a plain chocolate

4 Here are the nets of two different dice.

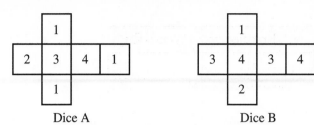

Dice A   Dice B

If both dice are rolled what is the probability that **(i)** using dice A
(ii) using dice B you will get:
(a) a 1   **(b)** a 4
(c) an even number   **(d)** an odd number
(e) a prime number   **(f)** a square number
(g) a number greater than 1   **(h)** a number less than 3?

5 Rashid spins the spinner. Each number is equally likely. What is
the probability he will get:

(a) a one   **(b)** *not* a one
(c) an even number   **(d)** *not* an even number
(e) a number greater than 2   **(f)** *not* a number greater than 2

Exercise 23.3   Links: (*23E*) 23E

1 (a) Toss a coin 50 times. Record your results in a table like this.
Work out from your results the probability of getting head
or a tail.

Coin	Tally	Frequency	Probability
Head			$\overline{50}$
Tail			$\overline{50}$
		Total	$\overline{50}$

Use your table to answer these questions:
(a) What is the probability of tossing a head?
(b) What is the probability of tossing a tail?
(c) What is the total of all the probabilities?
(d) Explain your result to part **(c)**.

2 Toss a coin another 50 times and complete a table as in
question **1**. Compare your results with question **1** and make
comments.

3 Toss two different coins 50 times. Record your results in a table like this:

Coins	Tally	Frequency	Probability
HH			
HT or TH			
TT			

(a) What is the probability of HH?
(b) What is the probability of HT or TH?
(c) What is the probability of TT?
(d) What is the total of the probabilities?
(e) What is the probability you will not get a TT?

Exercise 23.4   Links: (*23F*) 23F

1 This sample space diagram shows the outcomes when a coin is tossed and a dice is rolled.
Find the probability of:
(a) H1
(b) head and odd number
(c) tail and a 4
(d) tail or head and a 3
(e) tail or head and an even number
(f) tail and a prime number
(g) tail or head and a prime number

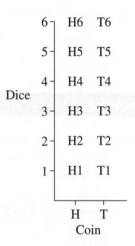

2 The spinner is spun once and the colour recorded. The spinner is spun a second time and the colour recorded.
(a) Draw a sample space diagram to show all the outcomes.
(b) List all the outcomes.

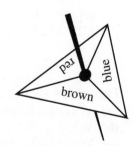

3 From a normal pack of playing cards a card is chosen, its suit recorded and then it is replaced. A second card is chosen, its suit recorded and then it is replaced.
(a) Draw a sample space diagram to show all the outcomes.
(b) List all the outcomes.

4 John is 3 times as likely to turn left at a T-junction as turn right. List all the possible outcomes after he has passed two junctions.

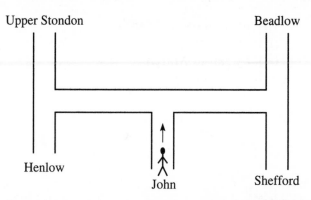

Upper Stondon   Beadlow

Henlow

John   Shefford

Exercise 23.5   Links: (23G) 23G

1 The number of students that used the computer room before school, at break and at lunch time is shown in this table. A student can only use the computer room once a week.

	Monday	Tuesday	Wednesday	Thursday	Friday	Totals
Before school	18	12		9	14	69
Break	33	25	28	22	31	
Lunchtime		41	37	34	43	
Total	85		81			

(a) Copy and complete the table.
(b) How many students used the computer room on Wednesday at break?
(c) How many students used the computer room on Thursday?
(d) How many students used the computer room before school during the week?
(e) If a student is chosen at random what is the probability that they used the computer room during:
   (i) Monday break   **(ii)** Wednesday lunchtime
   (iii) Friday   **(iv)** lunchtime in the week?

2 Mary and William asked 100 customers how they travelled to Edinburgh and in which month they travelled. Some of the information is recorded in this table.

	Coach	Rail	Car	Air	Totals
January	6		17	3	30
February		11	7	5	45
March					
Total		28	32	11	100

(a) Copy and complete the table.

(b) A customer is selected at random. What is the probability they travelled:

   (i) in February   **(ii)** in January or March

 (iii) by car   **(iv)** by car or air

   (v) not by coach   **(vi)** by rail in March

 (vii) by air in March   **(viii)** in January not by rail?

Note: There are no Practice exercises for
          Unit 24: Calculators and computers.

25 Scatter diagrams

1 Pupils' marks in a maths paper and a science paper were:

Maths	25	37	69	43	80	74	56	29	48	59	62
Science	31	38	66	47	76	79	58	30	54	56	65

(a) Draw a scatter diagram to represent this data.
(b) What type of correlation do you find?

2 What type of correlation, if any, would you expect if you compared the following data. Explain your answers.
(a) Shoe sizes and heights of people.
(b) The size of cars' engines and their highest speeds.
(c) The number of bedrooms in houses and the number of people living in them.
(d) The length of people's toe nails and their age.
(e) Students test scores in French and English.
(f) A person's income and their number of pets.

3 The handspan and the length of their little finger was measured for 10 students.

Handspan (mm)	201	182	174	207	203	191	197	184	189	171
Little finger (mm)	64	58	51	68	70	59	63	59	62	56

(a) Draw a scatter diagram to represent this data.
(b) What type of correlation do you find?
(c) Draw and label the line of best fit on your scatter diagram.
(d) Use your line of best fit to predict the finger length for a person whose handspan is 190 mm.

4 The height and shoe size of 12 people were measured.

Height	170	167	177	174	169	161	172	166	180	158	182	179
Shoe size	7	8	10	9	8	7	8	6	12	5	11	10

(a) Draw a scatter graph to represent this data.
(b) What type of correlation do you find?
(c) Draw and label the line of best fit on your scatter diagram.
(d) Use your line of best fit to estimate the length of a person whose shoe size is $7\frac{1}{2}$.

Heinemann Educational Publishers
Halley Court, Jordan Hill, Oxford, OX2 8EJ
a division of Reed Educational & Professional Publishing Ltd
Heinemann is a registered trademark of Reed Educational & Professional
Publishing Ltd

OXFORD MELBOURNE AUCKLAND
JOHANNESBURG BLANTYRE GABARONE
IBADAN PORTSMOUTH (NH) USA CHICAGO

Gareth Cole, David Kent, Peter Jolly, Keith Pledger, 1998, 2002

First published 2002

ISBN 0 435 53264 2

06 05 04 03 02
10 9 8 7 6 5 4 3 2 1

Designed and typeset by Techset, Tyne and Wear

Cover design by Miller, Craig and Cocking

Printed and bound by The Bath Press, Bath

Acknowledgements

The publisher's and authors' thanks are due to Edexcel for permission to reproduce
questions from past examination papers. These are marked with an [E]. The
answers have been provided by the authors and are not the responsibility of
Edexcel.